Synthetic Biology Conference 2014 (SEED 2014)

Engineering, Evolution & Design

Manhattan Beach, California, USA
14-17 July 2014

ISBN: 978-1-5108-1763-0

Printed from e-media with permission by:

Curran Associates, Inc.
57 Morehouse Lane
Red Hook, NY 12571

CURRAN ASSOCIATES INC.
proceedings
.com

Some format issues inherent in the e-media version may also appear in this print version.

Printed by Curran Associates, Inc. (2016)

AIChE
120 Wall Street, FL 23
New York, NY 10005-4020

Phone: (800) 242-4363
Fax: (203) 775-5177

www.aiche.org

Additional copies of this publication are available from:

Curran Associates, Inc.
57 Morehouse Lane
Red Hook, NY 12571 USA
Phone: 845-758-0400
Fax: 845-758-2634
Email: curran@proceedings.com
Web: www.proceedings.com

TABLE OF CONTENTS

STUDENT SESSION

AGRICULTURE AND PLANT ENGINEERING

RAPID DEBUGGING

MATERIALS AND NANOTECHNOLOGY

MAMMALIAN CELL ENGINEERING

ENGINEERING METABOLISM

COMPUTATION TO PHENOTYPE

NATURAL PRODUCTS: ACCESS AND DIVERSIFICATION

SYNTHETIC MICROBIOME

POSTER SESSION

Giving New Life to Materials for Energy, the Environment and Medicine

Belcher, A. M., Massachusetts Institute of Technology
Originally presented on: 7/14/2014 10:00:00 - 10:25:00

Organisms have been making exquisite inorganic materials for over 500 million years. These materials have many desired physical properties such as strength and regularity, which permit the organism to thrive in specific biological and physical environments. My lab seeks to expand the types of materials that living systems can utilize to make advanced technologies that are smarter and better adapted, using environmentally suitable techniques. One approach to designing future technologies that integrates the properties that living organisms use so well, is to select or evolve organisms to work with a more diverse set of building blocks. New nanostructured materials can be grown and assembled for energy storage, solar, carbon capture and re-use, catalysis, oil recovery and medical imaging by using genetic control and biologically inspired synthesis. In the field of energy storage, we demonstrate that biological engineering can enhance electrode design to improve specific capacity and cycling performance of both lithium-oxygen and lithium-ion batteries using M13 viruses to grow and assemble high-efficiency nanocatalysts from earth-abundant elements. In addition, rationally designing virus-based nanocomposites with high electron mobility can produce hybrid devices that can efficiently collect photo-generated electrons and greatly improves the performance of photovoltaic devices. M13 viruses can be further engineered as a scaffold to increase light collection, exciton transport and enable light driven reactions. Other applications include engineering yeast to convert carbon dioxide into building materials and the development of targeted probes in the second window near IR for detection and real-time surgical guidance of submillimeter ovarian tumors will be explored.

Nature, 405, 665, 2000

Science, 324,1051, 2009

Nature Communications 4, Article number: 2756, 2013

Nature Nanotechnology 6, 377, 2011

Energy Environ. Sci., 6, 660, 2013

Nano Lett.12 (3), 1176, 2012

Synthetic gene networks: emerging biotech applications

Collins, J. J., Harvard University

Originally presented on: 7/14/2014 10:25:00 - 10:50:00

Synthetic biology is bringing together engineers, physicists and biologists to model, design and construct biological circuits out of proteins, genes and other bits of DNA, and to use these circuits to rewire and reprogram organisms. These re-engineered organisms are going to change our lives in the coming years, leading to cheaper drugs, rapid diagnostic tests, and targeted therapies to attack "superbugs". In this talk, we highlight recent efforts to create novel synthetic gene networks and programmable cells, and discuss emerging synthetic biology applications in biotechnology and biomedicine.

ME models: their formulation and applications

Palsson, B., University of California - San Diego

Originally presented on: 7/14/2014 10:50:00 - 11:15:00

Genome-scale models of metabolism (M) and expression (E) in bacteria have been formulated. A ME model computes 80% of the proteome by mass, all RNA molecules, lipids, cell walls and the DNA. The non-ME proteome carries many creative functions some of which can be added into the ME, such as ATP maintenance functions and protein translocation processes. SUch a ME model provides a context for the expression of synthetic constructs and can compute explicitly how they tax the hosts mass and energy resources.

Direct Mutagenesis of Thousands of Genomic Targets using Microarray-derived Oligonucleotides

Bonde, M. T., Technical University of Denmark
Kosuri, S., Harvard University
Sarup-Lytzen, K., Technical University of Denmark
Church, G. M., Harvard Medical School
Wang, H. H., Columbia University
Originally presented on: 7/14/2014 13:30:00 - 13:55:00

Multiplex Automated Genome Engineering (MAGE) allows simultaneous mutagenesis of multiple target sites in bacterial genomes using short oligonucleotides. However, large-scale mutagenesis requires hundreds to thousands of unique oligos, which are costly to synthesize and impossible to scale-up by traditional phosphoramidite column-based approaches. Here, we describe a novel method to amplify oligos from microarray chips for direct use in MAGE to perturb thousands of genomic sites simultaneously. We demonstrated the feasibility of largescale mutagenesis by inserting T7 promoters upstream of 2587 operons in E. coli using this method, which we call Microarray-Oligonucleotide (MO)-MAGE. The resulting mutant library was characterized by high-throughput sequencing to show that all attempted insertions were estimated to have occurred at an average frequency of 0.02 % per loci with 0.4 average insertions per cell. MO-MAGE enables cost-effective large-scale targeted genome engineering that should be useful for a variety of applications in synthetic biology and metabolic engineering.

Genome Engineering Technologies, Strategies and Applications

Gill, R. T., University of Colorado

Originally presented on: 7/14/2014 13:55:00 - 14:20:00

The era of genome engineering well underway. DNA synthesis and sequencing technologies have now advanced to the point that they are no longer the rate limiting step in pathway to genome scale engineering projects. These technologies were used in the creation of the first synthetic genome. Such genome-construction technology was first applied to the copying of existing genomes. Current and future applications seek to design genomes that require increasingly sophisticated design capabilities, with a focus on engineering industrially relevant functions; such as production of biofuels, commodity or speciality chemicals, pharmaceuticals, industrial enzymes, etc. Such applications will require that we are able to not only identify genes encoding functions that enable such applications but also combinations of such genes, and combinations of such combinations, that together result in optimal organism performance. We are developing a range of new technologies for designing genomes based upon the i) construction and use of "ideal" chassis strains, ii) efficient identification of gene-to-phenotype design rules, iii) automated "rational" combinatorial mutation library generation, and iv) parallel interrogation of such libraries to identify combinatorial design rules. This presentation will describe the first generation of such technologies, the current state of next-generation approaches, and the most recent application of such tools to design and construct microbial genomes relevant to sustainable fuels and chemicals production.

Generation of human artificial chromosomes in human embryonic stem and neuronal cells

Monaco, Z.,

Originally presented on: 7/14/2014 14:20:00 - 14:45:00

De novo human artificial chromosomes (HAC) are non-integrating, high capacity vectors which function as stable normal chromosomes, and can incorporate large genes and the surrounding regulatory regions. HAC are composed of euchromatin and heterochromatin, and are models for investigating chromosome organization and gene expression. HAC are generated following transfer of large arrays of core centromeric DNA (alpha satellite DNA) into target cells. We developed a highly efficient delivery protocol based on the Herpes Simplex Virus type 1 (HSV-1) amplicon mediated transduction of cells, which targeted large DNA to the cell nucleus of different cell types. Recently HAC were stably generated at high frequency in human embryonic stem cells (hESc). Unlike tumour cells, where the input DNA often integrated in the host chromosomes, no integrated DNA was found in hESc. The HAC hES cells retained pluripotency and teratoma formation capabilities and were successfully differentiated into neuronal cells.

Development of the selection tools for the evolutionary engineering of genetic networks and for parallel and continuous editing of genomic DNA

Umeno, D., Japan Science and Technology Agency (JST)
Originally presented on: 7/14/2014 14:45:00 - 15:10:00

Directed evolution is a powerful strategy to construct complex yet functional systems ranging from genetic switch, regulatory circuits, and metabolic networks. To facilitate and accelerate the directed evolution of genetic switches and their assembly, we developed and tested various types of positive/negative selection systems for improving switching stringency, PoPS values, and rapidity in on/off transition. These methods were also adapted to the genome engineering, enabling the parallel, continuous, and seamless rewriting of multiple locus of bacterial genome.

1. Evolving for stringent switches and circuits
Reliable genetic networks require stringency in switch components. We developed a novel selection system using a viral nucleoside kinase as a single-gene dual selector. A single round of ON/OFF selection, all conducted by liquid handling, allowed us to enrich genetic circuitry with desired specification from variant pools, by the factor of 30,000x. Due to this power of the OFF selection, we could have isolated stringent circuits from the mixture containing non-stringent (leaky) ones.

2. Evolving temporal behaviours of genetic circuits
Synthetic biologists have been creating various timing circuits such as oscillators, pulse-generators, and delay switches (timers). For enabling the evolutionary design of temporal behaviour of the genetic circuits, we have developed another system, where both ON and OFF selections complete within 15 minutes. Using this selection, we have successfully enriched the rapid switchers, pulsegenerating circuits with desired time-profiles, and so on from the variant pools of the slowswitchers. Due to the rapidity of the selection process, one can also conduct serial operation of ON/OFF selections with reasonable time resolution to evolve temporal behaviour of the genetic circuits.

3. A method for continuous, iterative, and parallel operation of genome engineering
For the direct editing of bacterial genome with single-base resolution, two-step recombination methods are widely used. Here, the target site is replaced with PCR-generated DNA cassette first, and the cassette is then replaced with exogenous genes. Here, I report the systematic effort to redesign the system to improve the robustness, throughput, and speed of this method to realize the seamless and automation-compliant platform for genome engineering. Creating/testing the chimeric selectors, duplication of selector genes, tuning of the expression level of individual selector genes enabled us to eliminate the emergence of false-positive clones during the process. Having established the repeatable workflow where all the steps can be rapidly and seamlessly operated by liquid handling, we demonstrate the multi-step and in-parallel operation of genome modifications to generates a diverse set of isoprenoid-overproducing *Escherichia coli* strains.

Engineered Gene Circuits: From Oscillators to Synchronized Clocks and Biopixels

Hasty, J., University of California at San Diego

Originally presented on: 7/14/2014 16:00:00 - 16:25:00

This talk will focus on the development of synthetic gene oscillators and their synchronization. I willrst describe an engineered intracellular oscillator that is fast, robust, and persistent, with tunable oscillatory periods as fast as 13 minutes. Experiments show remarkably robust and persistent oscillations in the designed circuit; almost every cell exhibits large-amplitude uorescence oscillations throughout each experiment. Theory reveals that the key design principles for constructing a robust oscillator are a small time delay in the negative feedback loop and enzymatic protein decay that functions as an "overloaded" queue. I will then describe intercellular coupling that is used to generate synchronized oscillations in a growing population of cells. Micro uidic devices tailored for cellular populations are used to demonstrate collective synchronization properties along with spatiotemporal waves occurring on millimeter scales. While quorum sensing proves to be a promising design strategy for reducing variability through coordination across a cellular population, the length scales are limited by the di usion time of the small molecule governing the intercellular communication. I will conclude with recent progress engineering the synchronization of thousands of oscillating colony "biopixels" over centimeter length scales.

Communication and collaboration in synthetic microbial consortia

Collins, C., Rensselaer Polytechnic Institute

Originally presented on: 7/14/2014 16:25:00 - 16:50:00

The development of synthetic microbial communities presents unique opportunities for carrying out complex tasks through a division of labor approach. Each organism may be optimized individually and, much like at the genetic circuit level, organisms may be used in different combinations to carry out different processes or tasks. One of the major challenges with respect to engineering increasingly complex synthetic microbial consortia is the development of new biological components that enable cell-cell communication, and enable coordinated, population level behaviors.To this end, we have developed a new set of transcriptional regulators and promoters based on the *esa* quorum-sensing system that can be used to turn gene expression on or off in response to a cell-cell communication signal. We have also developed a set of AND-gate promoters to demonstrate that EsaR and its target DNA sequence can be used to engineer new combinatorial promoters that respond to cell-cell communication and a second, exogenous input. To expand our ability to use mixtures of diverse microbial species, we have developed a synthetic communication pathway between a representative Gram-negative organism (*Escherichia coli*) and representative Gram-postitive organism (*Bacillus megaterium*). An acyl-homoserine lactone (AHL)-dependent system was adapted to send signals from *B. megaterium* to *E. coli*. Components of a peptide-dependent microbial signaling pathway were used to send signals from *E. coli* to *B. megaterium*. We anticipate that our communication system, when combined with strategies for fine-tuning ecological interactions, will be a key technology for the implementation of synthetic consortia for bioprocessing, metabolic engineering and other applications.

Controlling Cells through RNA Folding

Lucks, J. B., Cornell University

Originally presented on: 7/14/2014 16:50:00 - 17:15:00

Cells have an amazing ability to process information, make decisions, and change their state in response to changing environments. This ability is encoded within the cellular DNA genome, which is converted into RNA and protein molecules through the basic processes of gene expression. Among the many functions these RNAs and proteins perform is regulating their own expression. In fact RNAs are now known to regulate almost all aspects of gene expression, and play central roles in controlling some of life's most basic processes.

As with many biomolecules, RNA function is intimately related to its structure. Being a single-stranded polymer of nucleotides, RNA can fold back on itself to form structures that enable certain RNAs to block or allow gene expression processes. Central questions in biology and bioengineering are then: How do RNAs fold *inside* cells?, and How can we engineer these folds to *control* gene expression? In this talk I will present our work in addressing both of these questions. I will start by describing SHAPE-Seq, our technology that couples chemical probing with next-generation sequencing to characterize RNA structures on an 'omics' scale. I will also describe our recent innovations in using SHAPE-Seq to simultaneously characterize the structures and functions of RNA regulatory molecules as they exist inside cells. I will then present our work on engineering RNA molecules that control the basic process of RNA transcription, and on how our knowledge of the RNA structure/function relationship has allowed us to engineer an RNA regulatory mechanism not found in nature. This offers the intriguing possibility that RNAs could be used to construct genetic circuits, and I will describe how we have been able to engineer a versatile set of RNA gene regulators that can be wired together to create RNA-only genetic circuits with fascinating properties compared to their protein counterparts. I will conclude with thoughts about how our work is leading us to ask deep scientific questions about the fundamental processes of RNA folding and function, and the role of RNAs in nature.

Developing CRISPR Tools to Reprogram the Mammalian Genome

Qi, L. S.,

Originally presented on: 7/14/2014 17:15:00 - 17:40:00

The genetic reprogramming of mammalian cell fate and identity requires efficient tools for genome engineering. The bacterial immune system CRISPR has emerged as a powerful approach for directed genome editing in diverse organisms via an RNA-guided Cas9 nuclease. In addition to its utility in genome editing, we have developed the CRISPR system as a versatile synthetic biology toolset for transcriptional regulation, epigenetic modification, and chromatin imaging in mammalian cells. We repurposed the catalytically inactive Cas9 (dCas9) that lacks the nuclease activity as synthetic transcription factors for sequence-specific activation or repression of endogenous genes. When coupled with RNA- or protein-interacting peptide domains, the system could be used for highly efficient, inducible, and multiplexable gene regulation. In sum, the CRISPR tools for gene regulation offer a new powerful approach for genetically reprogramming the mammalian cells and tissues.

Molecular programming with DNA/RNA

Yin, P., Harvard Medical School

Originally presented on: 7/14/2014 0:00:00 - 0:25:00

I will discuss my lab's research on engineering nucleic acid-based nanostructures and their applications in imaging, biosensing, nano-fabrication, and tissue engineering.

We have invented a general framework for programming the self-assembly of synthetic nucleic acid strands into prescribed target shapes or demonstrating their prescribed dynamic behavior. Using DNA, we have demonstrated the modular construction of sophisticated 1D (Science, 321:824, 2008), 2D (Nature, 485:623-626, 2012) and 3D (Science, 338:1177, 2012; Science, 334:65, 2014) structures on the 100-nanometer scale with nanometer precision. Using reconfigurable DNA hairpins, we have demonstrated diverse, dynamic behavior such as catalytic circuits, triggered assembly, and autonomous locomotion (Nature, 451:318, 2008). By interfacing these synthetic, nucleic acid nanostructures with functional molecules, we are developing diverse applications. In bioimaging, we have engineered geometrically encoded fluorescent barcodes for highly multiplexed single-molecule imaging (Nature Chemistry, 4:832-839, 2012) and dynamic fluorescent probes for highly multiplexed 3D super-resolution cellular imaging (Nature Methods, 11:313, 2014; Science, 334:65, 2014). In biosensing, we have constructed robust and ultraspecific probes for detecting single-base changes in singlestranded DNA/RNA targets (Nature Chemistry, 4:208-214, 2012). In nanofabrication, we have collaboratively developed a versatile framework for producing inorganic materials (e.g. graphene [Nature Communications, 4:1663, 2013], silicon dioxides [JACS, 135:6778, 2013], silver, gold) with arbitrarily prescribed nanometer scale shapes. In tissue engineering, we have developed a general strategy to engineer DNA directed self-assembly of biocompatible hydrogel bricks into complex architectures (Nature Communications, 4:2275, 2013).

I will also discuss engineering RNA nano-devices in living cells, in particular a RNA hairpin device that enables wide dynamic range and highly orthogonal regulation of gene expression and implements complex logic.

New Enzymes by Evolution: Expanding Nature's Catalytic Repertoire

Arnold, F. H., California Institute of Technology

Originally presented on: 7/15/2014 8:30:00 - 8:55:00

Synthetic biology for fuels and chemicals has depended on reassembling existing enzymes into new biosynthetic pathways. Unfortunately, many desired transformations fall outside the reach of known enzyme-catalyzed transformations or can be made more efficiently using synthetic chemistry. Thus, a future where metabolic engineering might produce nearly all of the organic molecules upon which society depends is still a ways off. Not satisfied with nature's vast catalytic repertoire, we are want to be able to create new enzymes to add to the toolkit of genetically encoded chemistry. My group uses the one proven algorithm for biological design—evolution—to optimize existing catalysts and create whole new ones. A powerful approach to engineering useful biological molecules, directed evolution both circumvents and underscores our profound ignorance of how sequence encodes catalytic function. I will describe various ways we have used evolution (contaminated with a little chemical intuition and design) to generate new catalysts starting from one of nature's most impressive, the cytochrome P450 monooxygenase. I will share several examples of engineered versions that catalyze important synthetic reactions not known in nature.

An Orthogonal Genetic System for Rapid Evolution

Liu, C. C., University of California at Irvine

Originally presented on: 7/15/2014 8:55:00 - 9:20:00

We recently developed an orthogonal DNA replication system in the yeast Saccharomyces cerevisiae.[1] This system consists of an orthogonal DNA plasmid–DNA polymerase pair wherein the orthogonal DNA polymerase stably and specifically replicates the orthogonal plasmid in the cytoplasm of yeast. Engineered error-prone DNA polymerases show complete mutational targeting *in vivo*: per-base mutation rates on the plasmid can be increased substantially with no increase in genomic rates. Orthogonal replication therefore serves as a platform for *in vivo* continuous evolution of user-selected genes and as a system whose replicative properties can be manipulated independently of the host's.

In this presentation, I will present the design and implementation of orthogonal DNA replication, discuss its use as a system for rapid targeted gene evolution *in vivo*, present new data on the addition of copy number control and recombination to our orthogonal replication system, and explore the broader implications of orthogonal replication as an unencumbered read/write system in biology that can be freely engineered for recording biological events *in vivo*.

[1]A Ravikumar, A Arrieta, CC Liu. An Orthogonal DNA Replication System in Yeast. *Nature Chemical Biology*, **10**, 175-177 (2014).

Ultrahigh-throughput Chemical and Biological Screening with Microfluidic Droplets

Abate, A. R., California Institute for Quantitative Biosciences, University of California, San Francisco

Originally presented on: 7/15/2014 9:20:00 - 9:45:00

Many important targets for directed evolution, like enzymes or biosynthetic pathways, yield molecules that are untethered to the proteins and genes responsible for producing them. Conventionally, evolving targets like these has been accomplished in microtiter plates or, when possible, in live cells via selections. While microtiter plates are amenable to assays that cannot be performed in cells, they are expensive and limited in the size of the libraries they can screen. Conversely, selections can screen large libraries, but are only possible with targets that can be assayed in living cells. We are developing an ultrahigh-throughput screening technology that combines the throughput of selections with the flexibility and generality of microtiter plate screening. In our approach, picoliter aqueous droplets used as "reactors" to assay individual members of the target libraries. Due to the small size of the droplets, and the ability to form, process, and sort them at kilohertz rates, we are able to screen libraries numbering in the hundreds of millions using microliters of total reagent, reducing cost and increasing throughput compared to microtiter plates by 1 million-fold. In addition, by integrating microfluidic screening with robotic fluid handling, we will be able to perform completely automated, continuous directed evolution of these targets.

GeneGuard: a Modular Plasmid System Designed for Biosafety

Ellis, T., Imperial College London

Originally presented on: 7/15/2014 11:00:00 - 11:25:00

Synthetic biology applications in biosensing, bioremediation and biomining envision use of engineered microbes beyond a contained laboratory. Deployment of such microbes in the environment raises concerns of unchecked cellular proliferation or unwanted spread of synthetic genes. To introduce security into bacterial synthetic biology, we have taken on the task of completely reformatting plasmids to be dependent on their intended host strain and inherently disadvantageous for others. Using conditional origins of replication, rich-media compatible auxotrophies, and toxin-antitoxin pairs we constructed a mutually-dependent host-plasmid platform, called GeneGuard. These devices, in isolation and in concert, severely reduce unintentional plasmid propagation in *E. coli* and *B. subtilis*, and do not disrupt their intended *E. coli* host's growth dynamics. Using a fluorescence-based assay we also show that they limit horizontal gene transfer from hosts to other bacteria and they work as efficient vectors for heavy-metal biosensors.

Engineering genetic sensors and circuits to construct programmable cells

Shopera, T., Washington University in St. Louis
Hoynes-O'Connor, A., Washington University in St. Louis
Immethun, C., Washington University in St. Louis
Henson, W. R., Washington University in St. Louis
Moon, T. S., Washington University in St. Louis
Originally presented on: 7/15/2014 11:25:00 - 11:50:00

The last decade has witnessed the tremendous power of synthetic biology in constructing simplified, synthetic genetic systems that can help understand complex biological processes. However, its full potential for practical applications has yet to be exploited. One of its promising applications is to build programmable cells which can integrate various environmental inputs and implement synthetic control over cellular processes. Here, we harness a toolbox of genetic parts (e.g., inducible promoters and regulatory proteins and RNAs) from a variety of organisms to build genetic devices and programs. To this end, such genetic parts have been mined from bacteria and assembled into genetic devices, which have been characterized using reporter proteins as outputs. Our programmable cells contain (i) environmental sensors, including oxygen, temperature, pH, and/or light sensors, which enable the engineered cells to distinguish different environmental conditions; and (ii) multi-input logic circuits with genetic memory modules, which direct the cells to generate outputs only when userdefined conditions are met. We will present progress on the creation of such engineered bacterial cells which can be used for real-world applications, mainly focusing on infectious disease prevention.

Host cell response to synthetic biology: a deeper look for more robust and optimised designs

Ceroni, F., Imperial College London
Garcia-Bellmunt, M., Imperial College London
Stan, G. B., Imperial College London
Ellis, T., Imperial College London
Originally presented on: 7/15/2014 13:30:00 - 13:55:00

Gene expression in living cells requires the use of host cell resources and precursors, like polymerases, ribosomes, amino acids and nucleotides. Synthetic biology systems thus place *de facto* a "load" on their host cells, which, in turn, causes a change in cell fitness and growth rate. Researchers are now becoming aware that this loading effect can lead to unpredictable interactions and unwanted behaviours that characterise many of the failures in synthetic biology, such as inactivating mutations.

Here, we present results pertaining to the characterisation of the burden caused by a library of synthetic biology devices embedded in two different *E. coli* strains. For this characterisation, a novel "capacity monitor" genetic device is used as a tool to quantify the burden imposed by synthetic gene expression via the measurement of both heterologous gene expression and host cell capacity with a standardised, plate-based assay that uses flow cytometry for single cell analysis.

Based on this method for burden quantification, we identify a variety of synthetic devices and pathways that cause significant burden to cells, and use quantitative PCR to reveal some native genes that significantly modulate their expression within the first hour of burden induction. These native genes represent potentially useful biomarkers of the native burden/stress. To understand and exploit this native burden response at a deeper level, transcriptome analysis of burdened cells is now in progress using RNAseq with the aim of identifying native regulators that can be repurposed in future burden-based feedback genetic devices.

Accelerating the design-build-test cycle of synthetic biological circuits in E. coli using S30 TX-TL cell-free systems, linear DNA, and modular assembly

Sun, Z. Z., California Institute of Technology
Noireaux, V., University of Minnesota
Murray, R. M., California Institute of Technology
Originally presented on: 7/15/2014 13:55:00 - 14:20:00

Decreasing the design-build-test cycle length is a fundamental challenge facing all engineering disciplines. This is acutely true in synthetic biology, where testing of circuits is limited by the lack of an intermediary, such as a breadboard, between the end chassis (a cell) and the starting engineering point. Limitations also exist in DNA assembly and cloning, as implementation of circuits *in vivo*requires all parts to be localized on plasmids with compatible origins of replications. While some limitations can be overcome by massively parallel and automated methods of cloning, transformation, and testing, the ability to do so is available only at select core facilities and is currently cost-prohibitive to widely emulate.

We propose using an open-source S30-based TX-TL cell-free system we have developed to serve as a "biological breadboard" intermediate for circuits before final *in vivo* implementation. Using protocols designed to require minimal infrastructure investment, the cell free system is extremely portable and generalizable. A typical design-build-test cycle length *in vivo* without automation can take a week or more. On plasmids in TX-TL, this length can be reduced to 1-2 days. Using linear DNA and a modular assembly platform, we demonstrate a further reduction to 4-8 hours. We show the implementation of short design-build-test cycles to prototype circuits such as genetic switches, novel RNA-regulated feed-forward loops based on repression and activation, and multiple protein-based feed forward loop variants. We also demonstrate the use of TX-TL to characterize robustness of individual synthetic biology "parts," and explore principles for transitioning results *in-vitro* to *in-vivo*.

Modular antibody-based protein sensors in mammalian cells for in-vivo rewiring of cellular fate

Siciliano, V., Massachusetts Institute of Technology
Beal, J., Raytheon BBN Technologies
Weiss, R., M.I.T.

Originally presented on: 7/15/2014 14:20:00 - 14:45:00

The potential of synthetic networks to reshape cellular behavior relies on the ability of engineered circuits to sense and process endogenous information and implement robust responses to intracellular conditions. We engineered a modular platform for programmable sensing-actuation devices in mammalian cells upon detection of a target protein. The genetic device couples the ability of single chain fragment antibody to specifically bind a target protein with TEV (Tobacco Etch Virus) protease activity. We successfully developed a programmable sensing-actuation device for proteins associated with hepatitis C virus (HCV), human immunodeficiency virus (HIV) and Huntington disease. Moreover, we demonstrate selective cell killing with incorporation of a pro-apoptotic gene, proving that our device has the potential to rewire cell fate. Our results demonstrate the capability of a modular framework to be adapted to detect different cellular states, and to reprogram cellular function by sensing protein biomarkers, in mammalian cells.

Deciphering global regulatory patterns for cellulase discovery in anaerobic fungi

Solomon, K. V., University of California, Santa Barbara
Henske, J. K., University of California, Santa Barbara
Haitjema, C., University of California, Santa Barbara
Borges-Rivera, D., Broad Institute of Massachusetts Institute of Technology and Harvard
Thompson, D., Broad Institute of MIT and Harvard
Regev, A., Broad Institute of Massachusetts Institute of Technology and Harvard
O'Malley, M. A., University of California, Santa Barbara
Originally presented on: 7/15/2014 14:45:00 - 15:10:00

The economical breakdown of plant biomass into simple sugars remains a significant challenge in the production of renewable cellulosic biofuels. Current strategies rely on cellulolytic enzyme cocktails from engineered aerobic fungi that are expensive to produce and inefficient against recalcitrant lignin-rich biomass. Thus, there is a critical need to develop new technologies to break down crude biomass into fermentable sugars for downstream fuel development. Towards this goal, much can be learned by studying how gut fungi depolymerize crude lignocellulose in the digestive tracts of large herbivores. These anaerobic fungi have evolved unique abilities to break down lignocellulosic biomass through invasive growth, and the secretion of powerful enzymes and synergistic mutli-enzyme complexes (cellulosomes). We have isolated a panel of unique gut fungi that thrive on lignocellulosic substrates, and secrete cellulosomal complexes of cellulases, hemicellulases, and cellulose binding domains that we are characterizing by next generation sequencing and proteomics. To determine the critical pathways of biomass hydrolysis within anaerobic fungi, we have used RNA-Seq to elucidate global regulatory patterns in response to catabolic repression of biomass degradation, and in response to growth on cellulosic substrates of increasing complexity. Through these efforts, we have identified more than a hundred transcripts encoding dozens of novel enzymes for biomass degradation. Many of these transcripts are strongly repressed by the addition of simple sugars and are clustered within distinct 'regulons' of coordinated gene expression. Gene set enrichment analysis confirms the upregulation of these regulons across cellulosic substrates and reveals how the composition of the fungal cellulosome adapts to various substrates. More importantly, the functional enrichment of these regulons suggests a critical role for the divergent unannotated transcripts that they contain, and points to a dozen interesting candidates for truly novel cellulose-degrading enzymes. Collectively, this information will establish the molecular framework for anaerobic fungal cellulose hydrolysis, ultimately allowing us to refactor this system in anaerobic fungi and beyond.

In- Situ Resource Utilization on Manned Martian Missions Using Synthetic Biology

Menezes, A. A.,
Cumbers, J.,
Hogan, J. A.,
Arkin, A. P., University of California at Berkeley
Originally presented on: 7/15/2014 15:10:00 - 15:35:00

Promising new technologies are currently sought by NASA for: generating fuels to propel primary spaceship and supplementary personal conveyances; manufacturing components in space; replacing cargoes of pre-packaged food; developing redundant life support systems; tackling diseases and increasing the shelf-life of therapeutics and pharmaceuticals; sensing and monitoring radiation levels and shield strength with light□weight apparatus; and recycling trash. Typically, the cost of these new technologies is reduced through *In Situ* Resource Utilization (ISRU), which consists of harnessing materials located at a mission's destination. Here, cost refers to launch mass, usable life, required power, or payload volume.

This work investigates how synthetic biology can meet several of NASA's needs. It reviews existing biological processes to demonstrate that they constitute a competitive yet non□traditional technology that is capable of processing volatiles and waste resources readily□available on a manned Martian mission in a way that reduces the launch mass of propellants, food and raw materials for 3D printing in space, and also overcomes the decreased product shelf□life of a common therapeutic. The work employs these reviewed processes in designs for natural and artificially□enhanced biological manufacturing strategies that can be leveraged to satisfy space input□availability and output□desirability constraints. It then analyzes methodological feasibility, technique versatility and the costs and yields of supporting apparatus and feedstocks, and compares potential future "space synthetic biology" advances to other new aerospace technologies.

When compared to non□biological approaches, this work determines that for 916□day Martian missions: 205 days of high□quality methane and oxygen Mars bioproduction with *Methanobacterium thermoautotrophicum* can reduce the mass of a Martian fuel□manufacture plant by 56%; 496 days of biomass generation with *Arthrospira platensis* and *Arthrospira maxima* on Mars can decrease the shipped wet□food mixed□menu mass for a Mars stay and a one□way voyage by 38%; 202 days of Mars polyhydroxybutyrate synthesis with *Cupriavidus necator* can lower the shipped mass to 3D print a 120 m^3 six□person habitat by 85%; and a few days of acetaminophen production with engineered *Synechocystis* sp. PCC 6803 can completely replenish expired or irradiated stocks of the pharmaceutical, thereby providing independence from unmanned resupply spacecraft that take up to 210 days to arrive. This proposed biological approach requires a few benign assumptions, such as complete nutrient extraction and recycling from the Martian soil; similar productivities in bioreactors on Earth and Mars; and an engineered acetaminophen productivity that is of the same order as engineered ethylene productivity when the organism chassis and the number of gene insertions are the same. The work recommends that near□term synthetic biology and related technological efforts focus on improving bioreactor nutrient recycling percentages, enhancing bioproduction of alternative nitrous oxide fuels, bettering flavors of *Spirulina*, testing interlocking 3D Printed polyhydroxybutyrate blocks in habitat and furniture construction, and increasing biosynthesis efficiencies of desired outputs, including pharmaceuticals.

Programmable multi-input transcription factor based circuit for the identification of specific cell types/physiological states

Angelici, B., ETH Zürich
Häfliger, B., ETH Zürich
Prochazka, L.,
Benenson, Y., Harvard University
Originally presented on: 7/15/2014 15:35:00 - 16:00:00

Confining a transgene activity to specific mammalian cell type, tissues or physiological state has been one of the major goals in genetic engineering with applications in basic research and gene therapy. The classic approach to restrict a transgene is to place it under the upstream region of an endogenous gene known for its specific expression in the tissue of interest. Many tissue- and cell type-specific promoters or enhancers have been used to this purpose, driving specific protein and microRNAs expression.

This approach ultimately relies on the existence of a natural promoter that is active only in the cell type or state of interest: while this assumption proves to be reasonable and successful in the case of specialized tissues or cell types, that need to express proteins with highly specific functions, usually doesn't hold for less specialized cells in a certain lineage or for pathological conditions. Pathological states in particular usually involve a dysfunctional protein or wide changes in expression patterns and rarely arise from (or result in) the organismal specific expression of a single gene.

Here we propose a scalable and programmable gene circuit able to sense transcription factors (TFs) activity in a modular fashion and integrating their levels, triggering a response only in case the inputs match a specific predetermined pattern. In particular we show a general strategy to sense, amplify and efficiently transduce the activity level of endogenous transcriptional activators. This robust transduction is then used to implement different Boolean logic functions.

As an initial proof of concept we apply this process to a small subset of TFs. Using a limited panel of NCI60 cells as a model system we show how a programmable logic integration of transcriptional regulators can be used to distinguish different lineage and potentially differentiate physiological and pathological profiles. Given the modular nature of the circuit this component can be used as a template to design similar sensor with different specificities and can be combined in larger pattern-sensing circuits using the same principles.

The specific conditional response of this system can be used for precise and specific manipulation of cell physiology: not only the final output can express an effector when all the criteria are met but each branch can be designed to interact with the cell through a miRNA or downstream transactivation in the presence of it's respective TF.

Bacteriophages use an expanded genetic code on evolutionary paths to higher fitness

Hammerling, M. J.,
Ellington, A. D., University of Texas at Austin
Barrick, J. E., University of Texas at Austin
Originally presented on: 7/15/2014 16:00:00 - 16:25:00

Despite the potential advantages of encoding protein sequences with an expanded alphabet that includes new amino acids with unique side chain properties, the genetic code consisting of the common 20 amino acids is nearly universally conserved in living organisms. Post-translational modifications augment the chemical repertoire of proteins in an ad hoc manner, but truly expanded genetic codes that enable direct ribosomal incorporation of a 21st amino acid into proteins are found rarely in nature and only partially reassign shared codons. Technologies exist to synthetically add a non-canonical amino acid (ncAA) to a genetic code by introducing an orthogonal aminoacyl-tRNA synthetase (aaRS) and a cognate tRNA recognizing the amber stop codon. To date, these systems have been too inefficient and genetically unstable to examine how the proteomes of organisms will evolve to adjust to and utilize a newly expanded genetic code. Here we show that T7 bacteriophages evolved on an *Escherichia coli* host that incorporates 3-iodotyrosine at amber codons became dependent on an alternative genetic code as they adapted to higher fitness. The evolutionary response included compensatory mutations that restored native protein termination and substitutions of amber codons in key viral genes. In particular, a 3-iodotyrosine substitution in the T7 type II holin protein was associated with more rapid lysis times and phage with this mutation were more fit than isogenic phage with canonical amino acids tolerated at this position. Thus, the expanded genetic code created new possibilities for adaptive mutations. It is now possible to thaw the "frozen accident" that resulted in a genetic code with the 20 common amino acids to examine how new chemical functionalities influence the evolutionary potential of proteins and entire organisms.

Reference:

Hammerling, M. J. et al. Bacteriophages use an expanded genetic code on evolutionary paths to
higher fitness. *Nat. Chem. Biol.* **10**, 178–80 (2014).

Engineered Microbial Swarmbots with Safeguard Mechanism

Huang, S., Duke University
Lee, A., Duke University
Pai, A., Duke University
Zhang, Y., Duke University
Leong, K., Duke University
You, L., Duke University
Originally presented on: 7/15/2014 16:25:00 - 16:50:00

Engineered bacteria have great potential for diverse applications in medicine, bioremediation and energy. These applications, however, have been limited by the safety concern over unanticipated effects of engineered bacteria on human health or environment. To address this issue, we have developed a prototype microbial swarmbots (MSB) technology to implement a safeguard mechanism and prevent unintended proliferation from controlled environment. We integrated synthetic biology, microencapsulation, and microfluidics to construct this technology. Specifically, the gene circuit we constructed is a minimal self-addition circuit to program density-dependent survival of the engineered bacteria: they can only survive at a sufficiently high density. When turned ON, the circuit will allow survival of encapsulated bacteria, which have a high local density, but not the escaped bacteria, which have a low density.
By encapsulating them in the microcapsules, the bacteria only grew inside the capsule not the outside as expected, when treated by appropriate concentration of antibiotic pulsing. Our mathematic models predicted all the MSB dynamics and enabled the optimal modulation in predictive manners. We further developed a microfluidic platform to monitor and quantify the dynamics, which confirmed the design features of MSB. Our work demonstrates a novel, integrated approach to achieve robust control of bacterial dynamics and has implications for autonomous synthesis and delivery of therapeutic proteins.

A Proposed Data Model for the Next Version of the Synthetic Biology Open Language

Roehner, N.,
Oberortner, E.,
Myers, C. J.,
Originally presented on: 7/15/2014 16:50:00 - 17:15:00

While the first version of the Synthetic Biology Open Language (SBOL) has been adopted by several academic and commercial genetic design automation (GDA) software tools, it covers a limited number of the requirements for a standardized exchange format for synthetic biology. In particular, SBOL Version 1.1 only allows the representation of hierarchically annotated DNA sequences of DNA components. This proposal extends SBOL Version 1.1 to include RNA and protein components, and non-genetic components such as small molecules and light. It also introduces modules, which instantiate groups of components that share molecular interactions and work together to perform a desired function. Taken as a whole, these extensions and other improvements to SBOL Version 1.1 enable the representation of a significantly larger portion of the information needed to exchange and reproduce a genetic design.

References:
Galdzicki, M. et al. Synthetic Biology Open Language (SBOL) Version 1.1.0. BBF RFC 87, 2012; DOI: 1721.1/73909.

A Strategy for Decoding Endogenous Plant Promoter Sequences

Megraw, M.,

Originally presented on: 7/16/2014 8:30:00 - 8:55:00

I will present a genome-scale dataset that precisely identifies Transcription Start Sites for a majority of Arabidopsis genes, revealing that plant promoters are not primarily TATA-based and have an unexpected structure composed of many position-specific sequence elements. A pattern recognition based analysis identifies combinations of factors that are likely to lead to transcription initiation. This analysis allows us to explore the question of what constitutes a 'minimal' endogenous plant promoter.

Evading Natural Selection in Complex Environments
Esvelt, K.,
Originally presented on: 7/16/2014 8:55:00 - 9:20:00

Most engineered genomic changes reduce evolutionary fitness outside the laboratory and are consequently eliminated by natural selection. By altering the environment, we can change the topology of the fitness landscape and select for altered traits such as herbicide resistance. However, these rationally designed solutions tend to be quite specific and are highly disruptive to existing ecosystems. The advent of CRISPR/Cas9 and other technologies that privilege design over evolution offer novel methods of evading natural selection with greater precision. These techniques may allow us to establish and maintain certain useful traits without otherwise perturbing the ecosystem.

Creating a Research Agenda for the Ecological Implications of Synthetic Biology

Collins, J.,
Kuiken, T.,
Oye, K., M.I.T.
Originally presented on: 7/16/2014 9:20:00 - 9:45:00

Synthetic biology is a field characterized by rapid rates of change and by the novelty, and breadth of applications. It is also an area of basic research and application that encompasses engineering along with the natural, physical, and social sciences. NSF's Division of Cellular and Molecular Biology, the Division of Environmental Biology, and the Engineering Directorate provided support for the MIT Program on Emerging Technologies and the Woodrow Wilson International Center for Scholars to organize two workshops to identify potential ecological effects of synthetic biology applications and the critical areas of uncertainty associated with those ecological effects. Further the project aimed to define technical research priorities to develop tools and methods to evaluate the ecological effects of synthetic biology applications and to identify the scientific research priorities to improve understandings of and ways to mitigate those ecological effects. Projects on nitrogen fixation, gene drive propagation in wild populations, glowing plants, and bio-mining were used to seed discussions among synthetic biologists, natural and social scientists and regulators. The consensus among workshop participants was that establishing and sustaining multidisciplinary research groups would be needed to address the following priority research areas: comparators; phenotypic characterization; fitness, genetic stability, and lateral gene transfer; control of organismal traits; monitoring and surveillance; modeling; and standardization of methods and data. Addressing these complex questions and overcoming communications barriers across disciplines cannot be done on a short term basis. In addition a concomitant assessment of economic and social implications of applications would be necessary to provide a foundation for assessing ecological implications. Synthetic biology is poised to make non-incremental, transformative advances in basic and applied areas of research. To realize these goals a coordinated prioritized research strategy amongst various governmental agencies, academia and industry will be needed to identify and mitigate the ecological implications of synthetic biology.

Implementation of cell-free biological networks at steady state

Niederholtmeyer, H.,
Maerkl, S. J.,
Originally presented on: 7/16/2014 11:00:00 - 11:25:00

Living cells maintain a steady state of biochemical reaction rates by exchanging energy and matter with the environment. These exchanges usually do not occur in in vitro systems, which consequently go to chemical equilibrium. This in turn has severely constrained the complexity of biological networks that can be implemented in vitro. We developed nanoliter-scale microfluidic reactors that exchange reagents at dilution rates matching those of dividing bacteria. In these reactors we achieved transcription and translation at steady state for 30 h and implemented diverse regulatory mechanisms on the transcriptional, translational, and posttranslational levels, including RNA polymerases, transcriptional repression, translational activation, and proteolysis. We constructed and implemented an in vitro genetic oscillator and mapped its phase diagram showing that steady-state conditions were necessary to produce oscillations. This reactor-based approach will allow testing of whether fundamental limits exist to in vitro network complexity.

Engineering synthetic ribosomes

Jewett, M., Northwestern University

Originally presented on: 7/16/2014 11:25:00 - 11:50:00

Purely *in vitro* ribosome synthesis could provide a critical step towards unraveling the systems biology of ribosome biogenesis, constructing minimal cells from defined components, and engineering ribosomes with new functions. To that end, we have developed an integrated synthesis, assembly, and translation technology (termed iSAT) for the *in vitro* construction of *Escherichia coli* ribosomes in crude ribosome-free S150 extracts. iSAT allows for the simultaneous *in vitro* synthesis of ribosomal RNA (rRNA), assembly of rRNA with purified ribosomal proteins, and transcription and translation of a reporter protein as a measure of ribosome activity. Here, we describe the development of, and recent improvements to, the iSAT system. Key breakthroughs include refining extract preparation, tuning the transcriptional balance of rRNA and mRNA production, and adding crowding and reducing agents. In addition to technological advances, we show that iSAT makes possible the *in vitro* construction of modified ribosomes by introducing a 23S rRNA mutation that mediates resistance against clindamycin. We also show that iSAT can be used for studying ribosome assembly. We anticipate that iSAT will aid studies of ribosome biogenesis and open new avenues for making ribosomes with altered properties.

Synthetic biology approaches to new nanomaterials

Chang, M. C., University of California, Berkeley

Originally presented on: 7/16/2014 13:30:00 - 13:55:00

Living systems have evolved the capacity to carry out many complex chemical transformations that could be potentially redesigned for the synthesis of target molecules of interest. One interesting example is the biogenesis of iron oxide nanocrystals in the directional-sensing bacterium, *Magnetospirillum magneticum* AMB-1. Magnetotactic bacteria have evolved a sophisticated system to sense the earth's magnetic field using linear chains of biological magnets, consisting of membrane-encapsulated nanoparticles built from magnetite (Fe_3O_4). These materials are biologically templated, producing defectless nanocrystals with a narrow size and shape distribution. Our goals in this area are to elucidate the biological mechanisms of iron mineralization by magnetotactic bacteria with the overall goal of developing synthetic biology approaches for the production of new nanomaterials using the magnetosome compartment.

Engineering Living Functional Materials

Chen, A. Y., M.I.T.
Zhong, C., M.I.T.
Lu, T. K., Massachusetts Institute of Technology
Originally presented on: 7/16/2014 13:55:00 - 14:20:00

Natural materials, such as bone, integrate living cells composed of organic molecules together with inorganic components. This enables combinations of functionalities, such as mechanical strength and the ability to regenerate and remodel, that are not present in existing synthetic materials. Taking a cue from nature, we propose that engineered 'living functional materials' and 'living materials-synthesis platforms', which incorporate both living systems and inorganic components, could transform the functionality and the manufacturing of materials. As a proof-of-concept, we recently demonstrated that synthetic gene circuits in *Escherichia coli* enabled biofilms to be both a functional material in its own right and a materials-synthesis platform. To demonstrate the former, we engineered *E. coli* biofilms into a chemical inducer-responsive electrical switch. To demonstrate the latter, we engineered *E. coli* biofilms into a synthesis platform that could organize biotic-abiotic materials across multiple length scales, template gold nanorods, gold nanowires, and metal/semiconductor heterostructures, and synthesize semiconductor nanoparticles[1,2]. Thus, tools from synthetic biology, such as those for artificial gene regulation, can be used to engineer the characteristic properties of living systems and to interface living systems with inorganic materials. Such hybrids possess novel properties enabled by living systems while retaining desirable functionalities of inorganic materials. These systems, as living functional materials and as living materials-synthesis platforms, provide a radically different paradigm of materials function and synthesis: materials possessing multifunctional, self-healing, adaptable, and evolvable properties that are synthesized in a distributed, bottom-up, autonomous, and environmentally sustainable manner.

References:
1. Chen, A. Y. *et al.* Synthesis and patterning of tunable multiscale materials with engineered cells. *Nat Mater* (Forthcoming 2014).
2. Chen, A. Y., Seker, U. O. S., Lu, M. Y., Citorik, R. J. & Lu, T. Synthesizing and Patterning Tunable Multiscale Materials with Engineered Biofilms. *bioRxiv*, doi:10.1101/002659 (2014).

Synthetic digital computation in human cells

Wong, W.,
Weinberg, B. H.,
Originally presented on: 7/16/2014 20:30:00 - 20:55:00

DNA recombinase, such as Cre, is one of the most powerful and widely used tools for regulating gene expression in mammalian systems. Currently, most works utilize only one recombinase, thus displaying limited spatiotemporal control. Here, through the use of orthogonal recombinases, we have developed a suite of synthetic genetic circuits with advanced computation capabilities in human cells. All 16 possible 2-input/1-output logic gates have been generated. In addition, multi-input/multi-output logic gates, such as half-adder, half-subtractor, and full-adder, have also been created. To demonstrate the multiplexing capability of DNA recombinases, we have constructed up to 8-input AND gates. Through the unique properties of DNA recombinase, we are able to significantly reduce the complexity of all the circuits and engineer the output generating components (non recombinase expressing portion), into a single transcription unit. These circuits pave the foundation for engineering advance computation in mammalian cells with direct applications in generating more accurate animal models and improving cell-based therapy.

CRISPR transcriptional repression devices and layered circuits in mammalian cells

Kiani, S., M.I.T.
Beal, J., Raytheon BBN Technologies
Ebrahimkhani, M. R., M.I.T.
Huh, J., M.I.T.
Hall, R. N., M.I.T.
Xie, Z., Tsinghua University
Li, Y., M.I.T.
Weiss, R., M.I.T.
Originally presented on: 7/16/2014 20:55:00 - 21:20:00

Engineered biological circuits offer insights into underlying biology of living cells and potential solutions to a range of medical and industrial challenges. To date, however, an impediment to engineering larger and more complex circuits in any living organism is the lack of an efficient framework for generating a sufficient number of composable devices. Recently, the Clustered Regularly Interspaced Palindromic Repeats (CRISPR) system has been adapted for both RNA-guided genome editing and gene regulation in a variety of organisms. CRISPR system provides new promises for generation of the transcription frameworks for diverse device library as it can be easily programmed by altering only the guide RNA (gRNA) sequence that complexes with Cas9 and targets the complementary DNA sequences. Therefore, we generated CRISPR-based devices in human cells that operate by targeting catalytically inactive Cas9(Ca9sm) and gRNA to a specifically designed hybrid promoter architecture that we call CRISPR responsive promoters (CRP). CRPs contain CRISPR target sites flanking mini-CMV promoter and upstream Gal4VP16 target sites allowing both activation by Gal4VP16 and repression by CRISPR targeting. In the presence of Cas9m and gRNAs complementary to target sites at CRP, the recruitment of the complex to CRP, represses an enhanced yellow fluorescent protein (EYFP) expression by steric blocking of the assembly of initiation complex. Examining such devices in human embryonic kidney (HEK) cells demonstrated up to 100-fold repression of EYFP output. Next, we examined the ability to create layered CRISPR-based circuitry in human cells. For such purpose, we rendered U6 promoters that drive gRNA expression repressible by CRISPR system and call the new promoter (CRU6). We then created a cascaded circuitry by connecting two of such devices. First, a gRNA-b will be expressed from a CRa-U6 promoter. Subsequently, the Cas9m and gRNA-b recruitment to CRP-b promoter represses EYFP expression. In the presence of an upstream device that contains a gRNA-a expressed from U6 promoter, targeting of Cas9m-gRNA-a to CRa-U6 promoter represses gRNAb expression and removes the repression on EYFP expression. By examining such circuit in HEK293 cells, we could achieve a functional cascade and derepression in EYFP output. In summary, we provide the first evidence that Cas9 system can be layered to generate circuitry of interconnected devices in human cells and propose a novel architecture to generate a large library of transcriptional devices for such purposes.

Circuits of Unknown Function

Elowitz, M.,

Originally presented on: 7/16/2014 21:20:00 - 21:45:00

Cells use circuits of interacting genes and proteins to control their behaviors, but the design principles that govern genetic circuit architecture usually remain mysterious. In order to figure out these design principles we combine three synergistic approaches: First, we apply time-lapse movies and quantitative image analysis to follow the dynamics of genetic circuits in individual cells. Second, we develop synthetic biology approaches in which we design and construct new genetic circuits (or re-wire natural ones) to implement specific functions in cells. Finally, we use mathematical models to explore the dynamics of actual and potential genetic circuit architectures. We work in bacteria, yeast, and mammalian cells, model systems that enable precise genetic manipulation and measurement of the dynamics of cellular gene circuits. In this talk, I will discuss recent work on design principles of the mammalian Notch signaling pathway.

Development of next-generation tools for efficient, costeffective, high-throughput engineering of yeast

Stanton, L.,
Chandran, S., Amyris
Hernday, A., Amyris
Originally presented on: 7/17/2014 8:30:00 - 8:55:00

Commercial scale microbial biosynthesis of chemicals requires repeating the design-build-test (D-B-T) engineering cycle multiple times in order to obtain a production-ready host. The timeline for the D-B-T engineering cycle directly affects the time to market for any new product. At Amyris, we have built the Automated Strain Engineering (ASE) pipeline that accelerates the D-B-T cycle, while simultaneous increasing the throughput of rational strain construction. While the ASE platform has succeeded at lowering the threshold in terms of cost and effort for building and testing strains, there is potential for further process improvements. To enhance the ASE pipeline and to enable the synthetic biology community make faster strain engineering progress, Amyris has continued to develop new molecular biology tools. In this presentation we will report the progress made on the building of some of these tools including:

i) A genome annotation pipeline capable of rapidly processing raw next-generation sequencing data for multiple genomes to standard parts

ii) A modified DNA assembly method capable of assembling large DNA constructs with high efficiency

iii) A next-generation sequencing pipeline for parallel sequencing of thousands of DNA assemblies at low cost

iv) Simultaneous multi-locus marker-free genomic integrations in various yeasts using nucleases

Widespread application of these tools will increase community access to high-throughput rational strain construction, resulting in the generation of Living Foundries at a rate, scale, and cost unprecedented in the field.

Exploring the potential of bacterial microcompartments for the spatial organization of synthetic metabolic pathways

Kim, E., University of California, Berkeley
Jakobson, C., University of California, Berkeley
Slininger, M.,
Tullman-Ercek, D., University of California, Berkeley
Originally presented on: 7/17/2014 8:55:00 - 9:20:00

Bacteria were long thought of as bags of enzymes with little internal structural organization. This idea has been challenged in recent years, particularly with the discovery of bacterial microcompartments (MCPs), which are polyhedral protein structures within the cytoplasm that house specific cellular reactions for functions such as carbon fixation. These MCPs may be a useful tool in the development of synthetic biology applications for which spatial control over the intracellular flow of proteins and small molecules is required. Toward this goal, we have been developing an understanding of bacterial MCPs and exploring the potential to engineer various features of the structures.

Bacterial MCPs consist of protein shells that sequester native enzymatic reactions and are thought to concentrate private cofactor pools, sequester toxic intermediates, and insulate the compartmentalized pathways from other cellular processes. Two well-characterized systems from *Salmonella enterica* use MCPs for 1,2-propanediol utilization (Pdu) and ethanolamine utilization (Eut). In both cases it is thought that the protein shell segregates the toxic aldehyde intermediate of the pathways from the cytosol, and that pores in the shell selectively mediate the passage of substrates and products into the shell lumen. In both the Pdu and the Eut MCP systems, short N-terminal extensions from native Pdu and Eut enzymes are sufficient to compartmentalize heterologous proteins in the protein shell. In order for these MCPs to be used for bioprocess applications, we must control the copy number, enzymatic loading and stoichiometry, size, and pore selectivity of these MCPs. To this end, we have used fluorescent reporters coupled with phenotypic and biochemical assays to study the regulation, copy number, and heterologous protein loading of the Pdu and Eut MCPs. We are simultaneously investigating the potential for controlling selectivity via shell protein pores, and have developed methods to heterologously express MCPs with tunable composition in *Escherichia coli*. Together, these experiments provide a framework and toolkit for the encapsulation of non-native pathways. Further efforts will enable the engineering of MCPs into tailored intracellular environments for applications such as the *in vivo* sequestration of synthetic metabolic pathways.

Identification of a Mono-Phosphomevalonate Decarboxylase and Utilization for Production of Isoprene without a Diphosphomevalonate Intermediate

Miller, M., DuPont Industrial Biosciences
Mampel, J.,
Primak, Y.,
Vaviline, D.,
Beck, Z.,
Whited, G., DuPont Industrial Biosciences
Originally presented on: 7/17/2014 9:20:00 - 9:45:00

Isoprenoids are secondary metabolites having value in a variety of commercial uses. At least three routes for the biosynthesis of isoprenoids have evolved, including two sequentially distinct routes from phosphomevalonate (MVP) to isopentenyl pyrophosphate in which the intermediate is phosphorylated and then decarboxylated (typical for eukaryotes) or decarboxylated and then phosphorylated (predicted for archaea). We have identified a MVP decarboxylase that completes the suite of enzymes needed for the predicted archaeal route. Isoprene production pathways following both the conventional eukaryotic route and this new route were used to bioconvert MVA to isoprene in *E.coli* at over 10 mg/L/hr/OD. This study demonstrates the value of applying synthetic biology not only to quantitative traits (*e.g.*, expression level and kinetic parameters) but also qualitative aspects (*e.g.*, chemical route) for pathway engineering. Such qualitative changes can be used to eliminate toxic heterologous metabolites, avoid interfering with endogenous pathways, and implement chemistries with improved fermentation performance.

Multiscale Models of Antibiotic Cellbots

Kaznessis, Y. N., University of Minnesota
Smadbeck, P., University of Minnesota - Twin Cities
Volzing, K., University of Minnesota
Borrero, J., University of Minnesota
Originally presented on: 7/17/2014 10:30:00 - 10:55:00

Foodborne gastrointestinal infections are significant causes of morbidity and mortality worldwide. Alarmingly, because of the extensive, non-therapeutic use of antibiotics in agriculture, foodborne bacteria are emerging that are resistant to our most potent drugs.

We will discuss a novel approach to reduce the use of antibiotics in food-producing animals and to treat gastrointestinal infections. We engineer lactic acid bacteria (LAB) that express and release antimicrobial peptides (AMPs). LAB are part of the gastrointestinal microflora and can be safely delivered with known benefits to humans and animals. AMPs are proteins that can be readily produced by LAB. One unique aspect of our approach is the use of synthetic promoters that precisely regulate the delivery of AMP molecules.

At the heart of proposed efforts are multiscale models that guide explanations and predictions of the antagonistic activity of recombinant LAB against pathogenic strains. Models are developed to quantify how AMPs kill bacteria at distinct but tied scales. Using atomistic simulations the various interaction steps between peptides and cell membranes are explored. Mesoscopic models are developed to study ion transport and depolarization of membranes treated with AMPs. Stochastic kinetic models are developed to quantify the strength of synthetic promoters and AMPs expression. We will also present a closure scheme for chemical master equations, providing a solution to a problem that has remained open for over seventy years.[1]

Experimentally, we engineer lactic acid bacteria to inducibly produce antimicrobial peptides. We have used a library of synthetic biological constructs.[2,3] We test these modified bacteria against pathogenic bacteria. We will present results against salmonella and enterococcus.[4]

1. Smadbeck P, Kaznessis YN. A closure scheme for chemical master equations. Proc Natl Acad Sci U S A. 2013 Aug 27;110(35):14261-5. doi: 10.1073/pnas.1306481110.
2. Ramalingam KI, Tomshine JR, Maynard JA, Kaznessis YN. Forward engineering of synthetic bio-logical AND gates, Biochemical Engineering, 47, 38-47, 2009.
3. Volzing K, Billiouris K, Kaznessis YN. proTeOn and proTeOff, New Protein Devices That Inducibly Activate Bacterial Gene Expression, ACS Chemical Biology, 6, 1107-1116, 2011.
4. Volzing K, Borrero J, Sadowsky MJ, Kaznessis YN. Antimicrobial Peptides Targeting Gramnegative Pathogens, Produced and Delivered by Lactic Acid Bacteria. ACS Synth Biol. 2013, 15;2(11):643-50. doi: 10.1021/sb4000367

Genetic coupling across space and time

Prindle, A., University of California, San Diego

Originally presented on: 7/17/2014 10:55:00 - 11:20:00

In many respects, bacteria are an ideal design platform for synthetic biology. They are small, extremely hardy, inexpensive to maintain, reproduce quickly, and comprise a vast array of species with unique properties. An expanding future landscape of *integrated* synthetic biology will use the native networks of diverse microbial species in concert with our own engineered gene circuits. In this talk, I will discuss two engineered gene circuits that incorporate host signaling machineries to communicate over large spaces and within short times.

I will first describe an LCD-like array of colony "biopixels" that collectively report the concentration of a target compound with a frequency-encoded signal[1]. These synchronous oscillations are maintained across populations as large as 13,000 colonies over centimeter length scales. Multi-scale coordination is achieved by layering 2 modes of communication – local quorum sensing and global redox signaling that utilizes the native aerobic response network.

I will then describe how we use protease competition to engineer rapid and tunable coupling of genetic circuits [2]. We characterize coupling delay times that are more than an order of magnitude faster than standard transcription-factor based coupling methods (less than one minute compared with □20-40 minutes). We use this mechanism as a platform to couple genetic clocks then show how the coupled clock network can be used to encode independent environmental inputs into a single time series output, thus enabling the possibility of frequency multiplexing in a genetic circuit context.

1. Prindle, A., et al., *A sensing array of radically coupled genetic 'biopixels'*. Nature, 2012. **481**(7379): p. 39-44.
2. Prindle, A., et al., *Rapid and tunable post-translational coupling of genetic circuits.* Nature (in press), 2014.

Automated metabolic pathway design and characterization

Anderson, J. C.,

Originally presented on: 7/17/2014 11:20:00 - 11:45:00

The automation and scaling of DNA synthesis and sequencing is ushering in a new era to metabolic engineering. Where once it was prohibitively costly to test many designs during a project, automated fabs are being built to prototype biosynthetic pathways and test them on a massive scale. In conjunction with these efforts, my lab has written algorithms to automate the design of biosynthetic pathways. The Act Ontology is an aggregator of standardized Observations of biochemical reactions. We have populated the Ontology from public databases and natural language processing (NLP) of PubMed abstracts. Upon this ontology we have written algorithms that process and abstract the observed reactions, to bin them by chemical-relatedness. Upon these abstractions we employ a design synthesis tool that uses these models to predict sets of genes that, when added to a particular cell, will result in the production of a target biochemical. The tool can exhaustively permute the chemical space accessible by concrete monofunctional enzyme reactions, which we refer to as the Reachable molecules. It can also extrapolate this space by including one speculated reaction. These Reachables can then be surveyed in various ways for interest in terms of societal benefit, profit, biosafety, regulatory, or intellectual property concerns. For any one of the Reachables, Act can compute all known paths to that molecule and all known genes that could satisfy each reaction in each path. Concrete predictions can thus be enumerated and then ranked for plausibility by incorporation of expression data, thermodynamic considerations, and the like. We have demonstrated the utility of the tool through the biosynthesis of Acetaminophen (Tylenol), an unnatural chemical that is surprisingly only one enzymatic step away from native E. coli metabolism. Additionally we have demonstrated unexpected glucaric acid pathways that included predictions mined through NLP methods. Ultimately, the algorithms are only as accurate, detailed, and extensive as the bag of Observations that support them. In parallel to these studies, my wetlab has developed multiplex characterization assays to scale the process of acquiring new Observations. These methodologies include MOLSET, a workflow from DNA microarrays to expression data acquired through deep sequencing. A similar technique, DLENCA allows monitoring of enzymatic reactions at scale. Large data sets from such experiments and pathway prototyping experiments are fed into Act as Observations. This completes an automated design-test-learn cycle where new scientific knowledge automatically accumulates through the testing of predictions.

Biocompatible Chemistry

Balskus, E.,

Originally presented on: 7/17/2014 13:30:00 - 13:55:00

Synthetic organic chemists and synthetic biologists currently use largely orthogonal technologies to access small molecules like pharmaceuticals and commodity chemicals. As the use of biological catalysts and engineered organisms for chemical production grows, it is becoming increasingly evident that future efforts for chemical manufacture will benefit from the integration and unified expansion of these two fields. An underexplored strategy for manipulating metabolic pathways and intermediates is the introduction of non-enzymatic reactions from organic synthesis into living systems. In this talk, I will describe efforts to develop biocompatible chemistry: non-enzymatic transformations that can be used to alter the structures of small molecules in the presence of living organisms. Interfacing such reactions directly with bacterial metabolism offers new opportunities to use biology for small molecule synthesis as well as novel approaches for manipulating the growth and behavior of living organisms.

Discovery of Novel Natural Products via Synthetic Biology

Zhao, H., University of Illinois at Urbana Champaign
Originally presented on: 7/17/2014 13:55:00 - 14:20:00

Microorganisms are a major source of new therapeutic agents. Sequenced genomes and metagenomes provide a tremendously rich source for discovery of novel gene clusters involved in natural product biosynthesis. However, due to the lack of tools to efficiently identify a complete biosynthetic gene cluster, determine the role of each involved gene, and subsequently designate a function to the target gene cluster, only a tiny fraction of those putative natural product biosynthetic gene clusters have been characterized (1). To overcome this limitation, my group has developed a new genomics-driven, synthetic biology enabled method to discover and produce novel natural products from sequenced genomes and metagenomes. By taking advantage of the highly efficient yeast *in vivo* homologous recombination mechanism, this method refactors the target cryptic pathway together with the genetic elements needed for DNA maintenance and replication in *S. cerevisiae*, *E. coli*, and a target heterologous host using a plug-and-play scaffold and a set of heterologous promoters that are functional in the target heterologous host. As proof of concept, we awakened the silent polyketide spectinabilin pathway from *Streptomyces spectabilis* in *Streptomyces lividans* (2) and activated a cryptic pathway containing a polyketide synthase-non-ribosomal peptide synthetases from *Streptomyces grieseus* in *Streptomyces lividans*, which led to the discovery of two novel tetramic acid natural products that have never been reported in literature (3). Our method bypasses the traditional laborious processes to elicit pathway expression and represents a new platform for discovering novel natural products (4). Currently, in collaboration with several research groups including but are not limited to Adam Arkin, James Allison, Bill Metcalf, and Neil Kelleher that have expertise in bioinformatics, genomics, mass spectrometry, and automation, we are establishing a fully integrated robotic system to automate all the steps in gene cluster refactoring and product detection, which should significantly increase the throughput of our method and discover novel pharmaceutically important natural products.

1. R. E. Cobb and H. Zhao. "Direct Cloning of Large Genomic Sequences." *Nature Biotechnology*, 30, 405-406 (2012).

2. Z. Shao, G. Rao, C. Li, Z. Abil, Y. Luo, and H. Zhao. "Refactoring the Silent Spectinabilin Biosynthetic Gene Cluster Using a Plug-and-Play Scaffold." *ACS Synthetic Biology*, 2, 662-669 (2013).

3. Y. Luo, H. Huang, J. Liang, M. Wang, L. Lu, Z. Shao, R. Cobb, and H. Zhao. "Activation and Characterization of a Cryptic Polycyclic Tetramate Macrolactam Biosynthetic Gene Cluster." *Nature Communications*, 4: 2894 (2013).

4. R. Cobb, J. Ning, and H. Zhao. "DNA Assembly Techniques for Next Generation Combinatorial Biosynthesis of Natural Products." *Journal of Industrial Microbiology and Biotechnology*, 41, 469-477 (2014).

The Potential and Challenges of Cryptic Metabolites
Clardy, J., Harvard Medical School
Originally presented on: 7/17/2014 14:20:00 - 14:45:00

Small molecules biosynthesized by environmental microbes – bacteria and fungi – have provided the molecular basis for the majority of therapeutic agents used as antibiotics, anticancer agents, and immune system modulators. In spite of the historical success of this discovery paradigm, it has been largely abandoned by the pharmaceutical industry for many reasons including high rediscovery rates of known compounds. Over the last decade, genomic sequences from both bacterial and fungal genomes have revealed that we typically discover ~10% of the genetic potential of producing organisms under standard laboratory conditions. The remaining 10% are called 'cryptic metabolites', and their biosynthetic genes are called 'orphan clusters'. Finding cryptic metabolites, or deorphanizing clusters, has been pursued by several approaches, which will be discussed in this talk. The failure of any robust algorithmic discovery paradigm to emerge from these efforts creates a significant opportunity for synthetic biology.

Digital Cell Division Counting Reveals Microbiotal Dynamics in the Mammalian Gut

Myhrvold, C., Harvard Medical School
Kotula, J., Harvard University
Hicks, W., Harvard Medical School
Silver, P. A., Harvard Medical School
Originally presented on: 7/17/2014 16:00:00 - 16:25:00

The gut microbiota is a vibrant ecosystem of critical importance to animal health. To study microbiotal dynamics, we engineered bacteria that count cell divisions using genetically encoded, digitally segregating fluorescent particles. We tested our system by counting for up to 14 generations in continuous culture across an order of magnitude in growth rate. We then engineered *E. coli* from the mouse gut, introduced them into mice, and determined that cells double every 3 hours during the first 12 hours of passage through the animal. Taken together, we have generated the first living digital counter that can function in animals and thus can be used to further study the physiology of microbes in diverse environments.

Parasight – A synthetic biology approach for the detection of parasites

Webb, A. J., Imperial College London
Kelwick, R., Imperial College London
McKeown, C., Imperial College London
Jensen, K., Imperial College London
Kylilis, N., Imperial College London
Baldwin, G. S., Imperial College London
Ellis, T., Imperial College London
Freemont, P. S., Imperial College London
Originally presented on: 7/17/2014 16:25:00 - 16:50:00

One of the primary goals of synthetic biology is the application-driven generation of new parts, circuits, and systems to solve problems that as yet have not been adequately addressed. The parasitic infection *Schistosomiasis* affects over 200 million people worldwide, with estimates suggesting that a further 780 million people are at risk of infection. The causative agents are fluke worms of the *Schistosoma* genus, and infection only occurs when the cercarial larvae are able to penetrate the skin. To facilitate this, the cercariae secrete an elastase possessing a defined substrate specificity. Our project takes advantage of this property of the cercarial elastase, and has used it to design and create biosensors that are specific in targeting *Schistosoma*. The design of our biosensors is based on a two-pronged approach, 1) an accurate detection system which is targeted to the cercarial elastase, and 2) an easily measurable output. Using synthetic biology approaches we have engineered the biosensors to be "housed" in two bacterial chassis, *Bacillus subtilis* and *Escherichia coli*. The *B. subtilis* biosensors comprise a fusion protein that possesses the cell wall binding domain of LytC along with our biosensor component, whilst the *E. coli* biosensors are composed of the CmpX protein, which traverses the outer membrane, fused with our biosensor component. Both biosensor components are presented to the external environment, and possess cercarial elastase recognition motifs, as well as either a streptavidin binding peptide for the *E. coli* biosensors, which can be detected by a streptavidin fluorophore conjugate, or a poly-histidine tag for the *B. subtilis* biosensors, which can be detected using a poly-histidine tag antibody fluorophore conjugate. Thus, when the elastase is present it cleaves the biosensor at the recognition motif site, thereby releasing the detectable components. Here we report and discuss the progress we have made in this important project and share some exciting data.

Probiotics engineered to fight against human pathogens

Hwang, I. Y., National University of Singapore
Koh, E.,
Wong, C. K., Nanyang Technological University
Lo, T. M.,
Chang, M. W., National University of Singapore
Originally presented on: 7/17/2014 16:50:00 - 17:15:00

Synthetic biology aims to engineer genetically modified biological systems that perform novel functions that do not exist in nature, with reusable, standard interchangeable biological parts. By having this engineering framework in place, synthetic biology has the potential to make the construction of novel biological systems a predictable, reliable, systematic process. In our previous study, we engineered *Escherichia coli* to effectively sense the presence of *Pseudomonas aeruginosa* and kill the pathogen by releasing antimicrobial peptide, as a proof-of-concept of a novel microbe-based antimicrobial strategy. In this study, we have chosen a probiotic strain, to develop into therapeutic microbes with clinically relevant functionalities to fight against human pathogen. The implemented functionalities include the ability of the probiotic microbes to swim toward *P. aeruginosa* upon sensing and disrupt biofilm, and exert strong antimicrobial activity against gastrointestinal infections. The *next-generation* engineered cells were able to demonstrate pathogen specific cell migration that can sense and kill the pathogen. Furthermore, this target localization was observed in the presence of mature biofilm, which is responsible for antibiotic-resistance and chronic infection, thereby exhibiting significantly improved killing efficiency compared to its non-motile counterpart. The effective antimicrobial activities against target human pathogens have also been shown in eukaryotic infection models. This development of pathogen targeting probiotics presents the potential use of a synthetic biology-driven antimicrobial strategy that can complement current antibiotic therapies to be a cost-effective therapeutic agent that is capable of rapid self-diagnosis and treatment against an infection.

Insights from a Global Analysis of Secondary Metabolism: Small Molecules from the Human Microbiome

Fischbach, M.,

Originally presented on: 7/17/2014 17:15:00 - 17:40:00

The discovery of natural products – small molecules from microbes often used as drugs – has been an ad hoc pursuit for almost a century. The rapidly growing database of microbial genome sequences offers new opportunities to leverage genomics and bioinformatics toward discovering natural products and characterizing their roles in mediating interspecies interactions. This seminar will describe two convergent, ongoing lines of research: our use of genomics and bioinformatics to identify biosynthetic gene clusters and predict the structures of their small molecule products, and our efforts to identify and characterize small molecules produced by the human microbiota.

Quantitative Design of Layered Genetic Circuits

Nielsen, A.,

Engineered bacteria have broad potential for biotechnology, including materials production, food and energy, industrial biosynthesis, and biosensing. Synthetic genetic circuits enable the rational programming of bacterial behavior to respond to specific environmental signals with conditional gene expression. To advance genetic circuit design capabilities, we have developed a forward- engineering approach that uses characterized transcriptional modules and quantitative modeling of circuit function. Using this approach, we have designed and implemented complex multi- input, multi-output genetic logic gates. These circuits demonstrate the recent strides made toward engineering novel cellular behaviors *de novo*.

Synthetic Chromatin-Based Logic, Spatial Regulation, and Memory
Keung, A., Howard Hughes Medical Institute
Bashor, C. J., Howard Hughes Medical Institute, Boston University
Kiriakov, S., Boston University
Collins, J. J., Howard Hughes Medical Institute, Boston University
Khalil, A. S., Boston University

Synthetic chromatin-based logic, spatial regulation, and memory

Albert J. Keung1,3,5, Caleb J. Bashor1,3, Szilvia Kiriakov2, James J. Collins1,3,4, Ahmad S. Khalil1,4,*

1Department of Biomedical Engineering and Center of Synthetic Biology, Boston University, Boston, MA 02215, USA

2Program in Molecular Biology, Cell Biology & Biochemistry, Boston University, Boston, MA,

02215, USA

3Howard Hughes Medical Institute

4Wyss Institute for Biologically Inspired Engineering, Harvard University, Boston, MA 02115, USA

5Presenter

*Correspondence: akhalil@bu.edu (A.S.K.)

ABSTRACT

The complex structure of chromatin vastly extends the information potential of the eukaryotic genome beyond the underlying DNA sequence. Several hundred chromatin regulators (CRs) modulate eukaryotic genome structure and function by remodeling nucleosomes and DNA. Yet, despite being the subject of extensive studies, it remains unclear how CRs operate through diverse processes to regulate gene transcription, and this powerful layer of gene regulation has yet to be exploited in engineering applications. Synthetic biology offers a unique approach to decompose and exploit the complexity of chromatin by studying emergent transcriptional behaviors from engineered combinatorial, spatial, and temporal patterns of individual CRs. Here we fuse a library of 223 yeast CRs to programmable zinc finger proteins. Site-specific and combinatorial recruitment of CRs to engineered genomic loci drives a host of transcriptional logic functions, spatial behaviors, and epigenetic control. Analyzing these transcriptional behaviors through gene ontology provides useful design principles for CR organization in establishing and engineering chromatin-based transcriptional regulation. This work presents a bottom-up approach to investigating chromatin regulation, and introduces new, chromatin-based components, useful transcriptional behaviors, and design strategies for synthetic biology and cellular engineering.

Mpath: Computationally-Aided Design of Synthetic Metabolic Pathways
Cox, R. S. III,
Nakatsui, M., Graduate School of Engineering, Kobe University
Makiguchi, H., Mitsui Knowledge Industry Co.
Ogawa, T., Mitsui Knowledge Industry Co.
Kondo, A., Graduate School of Engineering, Kobe University
Araki, M., Kobe University

Mpath: Computationally-aided design of synthetic metabolic pathways

Robert Sidney Cox III1 Masahiko Nakatsui1 Hiroki Makiguchi2 Teppei Ogawa2 Akihiko Kondo1

Michihiro Araki1

We introduce a computational platform, Mpath, for exploring synthetic metabolic pathways including putative enzyme and compound information. Mpath samples randomly combinations of known chemical reaction steps to identify potential metabolic routes between target compounds. We demonstrate this heuristic algorithm to design putative pathways for the production of novel synthetic amino acids *in vivo.*

Keywords â?? metabolic engineering, database, chemoinformatics

I. BACKGROUND:

HE growing catalogue of known enzymes promises new biological products by combining transgenes *in vivo* and engineering enzymes with novel catalytic capabilities. Amino acid production is part of the core metabolism of every free-living organism, but derivatives of the standard
20 amino acids are often produced as part of the â??secondary metabolismâ?? [1]. For example, many neurotransmitters are derivatives of standard amino acids, and their derivatives are rich targets for drug discovery.
In this study we have developed an efficient method

-
handling comprehensive enzymatic reaction data to design extensive metabolic pathways including putative compounds and enzymatic reactions. We first developed a simple

method to represent chemical structures in the form of feature vectors and enzymatic reactions using feature differences between chemical pairs. An algorithm was then developed to find possible combinations of chemicals and enzymatic reactions from start to target compounds on the basis of linear programming.

The design of metabolic pathways is finding combinations of reaction features that satisfy differences between two chemical features. The reaction features are rearranged to yield chemical features in sequence, which are used to assign compounds from our chemical database by similarity. Our method significantly reduces the computational time to find extensive metabolic pathways. The resulting metabolic pathways including putative compounds and enzymatic reactions are ranked on the basis of feasibility criteria using chemical similarity and stored in a pathway database. A web user interface is also developed to check pathway candidates.

1Organization of Science and Technology, Kobe University, 1-1

Rokkodai Nada, Kobe 657-8501, Japan

2Mitsui Knowledge Industry Co., Osaka Mitsui-Bussan Bldg. 6F, 2-3-33

Nakanoshima, Kita-ku, Osaka 530-0005, Japan

*Presenter E-mail: Sidney@dna.caltech.edu

II. RESULTS

A. We began with the rich set of more than 11,000 L- amino acid-like compounds in the PubChem database [3] and using these as target molecules, calculated acceptable metabolic pathways for their synthesis from glucose. For this we used the annotated enzymes from the KEGG database [2] and identified reaction steps between compounds that are currently present in nature, along with enzymatic steps which might be easily engineered due to high chemical similarity between known and target compounds. We scored each reaction step by chemical similarity. Mpath correctly reconstructed pathways for 50 amino acid derivatives which are contained in KEGG, but are not part of the core reference pathway.

B. From the 1,987 putative synthetic amino acid pathways, we analyzed 100 and chose the 50 most feasible for classification. Most commonly, the sidechains of lysine, glutamic acid, cysteine, and serine were found to participate in several 'sidechain linking' derivative reactions. Several small molecules were found to attach to these amino acids including carboxylic acids such as acetic acid, formic acid, succinic acid, and also small primary amines. Other pathways classifications included amino transfer reactions, catabolic degradation pathways, and aromatic ring substitutions.

III. CONCLUSION

The Mpath algorithm exhibits several useful properties compared to other methods, including speed when calculating synthetic pathways composed of many steps from large reaction databases. We found many putative pathways for making new amino acid derivatives, which might be useful for pharmaceutical screening and materials engineering applications. In particular we found a high variety of glutamic acid side-

chain linked compounds, and we present these for possible applications.
REFERENCES

[1] Akashi, H. and Gojobori, T. 2002. Metabolic efficiency and amino acid composition in the proteomes of Escherichia coli and Bacillus subtilis. *Proceedings of the National Academy of Sciences of the United States of America*. 99, 6 (Mar. 2002), 3695â??3700.

[2] Kanehisa, M. et al. 2010. KEGG for representation and analysis of molecular networks involving diseases and drugs. *Nucleic Acids*

Research. 38, Database issue (Jan. 2010), D355â??60.

[3] Wang, Y. et al. 2009. PubChem: a public information system for analyzing bioactivities of small molecules. *Nucleic Acids Research*.

37, Web Server issue (Jul. 2009), W623â??33.

Evolution of a Yeast Receptor for Non-Native Ligands Using a Novel High-Throughput Screening Method

Adeniran, A., Northwestern University
Bostick, J., Northwestern University
Pertusi, D., Northwestern University
Reddick, M.,
Sherer, M.,
Riley, T.,
Tyo, K. E. J., Northwestern University

This research aims to develop novel yeast biosensors that take advantage of the yeast pheromone mating pathway, one of the best understood eukaryotic signaling pathways. Through this pathway, *Saccharomyces cerevisiae* detects mating pheromone with the Ste2 G-protein coupled membrane receptor (GPCR). Though *S. cerevisiae* has been used as a system to express heterologous mutagenized GPCRs, this research takes advantage of the native signaling machinery in yeast and eliminates the need for extensive pathway engineering that is often necessary for functional expression of heterologous receptors. The objective of this experiment was to determine if the Ste2 receptor could be mutated to yield novel detection capabilities. Error-prone polymerase chain reaction was used to create a library of mutagenized Ste2 receptors. We searched this library for receptors that showed activity for variants of the native pheromone ligand. These variants differ from the native ligand by one amino acid, and the native Ste2 receptor shows little activity for these pheromone variants at physiological concentrations.

Our results indicate that we were successful in identifying mutated receptors that respond to the pheromone variants. The library of mutant Ste2 receptors was sorted with fluorescence activated cell sorting (FACS). Developing a sorting protocol based on Bayesian statistics allowed us to minimize the amounts of false positive and false negative receptors in the mutant population. Receptor activation was indicated by green fluorescent protein (GFP) expression. The GFP value was normalized by red fluorescent protein (RFP), which was constitutively expressed. After FACS, individual mutant receptors were isolated from the population for characterization. Dose response curves were obtained by treating individual receptors with varying amounts of pheromone variant. EC50 values, the concentration of ligand at which half of the maximal response is obtained, were calculated for each isolated receptor.

For mutant receptors, the pheromone variant EC50 values obtained were significantly lower than those obtained with the native Ste2 receptor, implying that the isolated receptors are more sensitive than the native Ste2 receptor is for the pheromone variants. These results lay the foundation for future work for the directed evolution of the Ste2 receptor to detect non-native ligands.

Engineering Auto-Regulatory Genetic Circuits That Enable Programmable Biological Functionalities: Biotechnological Applications

Lo, T. M., National University of Singapore
Chang, M. W., National University of Singapore

Engineering auto-regulatory genetic circuits that enable programmable biological functionalities: Biotechnological applications

Tat-Ming Lo and Matthew Wook Chang
Department of Biochemistry, Yong Loo Lin School of Medicine, National University of Singapore, Singapore 117456

Microbes can be reprogrammed to demonstrate biotechnologically important functionalities such as biochemicals production. However, reprogramming microbes to perform complex and coordinated tasks often requires the use of additional exogenous chemicals such inducers, which can increase production costs, introduce host toxicity and incompatibilities with industrial scale-up. To overcome this challenge, we developed auto-regulatory genetic circuits by introducing novel sensors and feedback loops into microbial systems. Not only our engineered systems are independent of costly inducers, they are capable of regulating protein expression based on desired features such as substrate concentration and cell-density. One example is the use of our genetic circuit in auto-regulatory lignocellulosic bioconversion. In this system, the genetic circuit was engineered to regulate enzyme expression based on lignocellulose substrate cues and cell-density. In another example, we used our cell density-dependent circuit to facilitate product extraction in *Escherichia coli*. Macromolecular product extraction in *E. coli* often requires chemical, enzymatic, and/or mechanical lysis, thus increasing production costs. To aid in the extraction of biochemical products and simplify the product harvest process, we developed a genetic circuit that enabled cell density-dependent auto- regulatory lysis based on a transducer-switch scheme. As a proof of concept, we demonstrated the applicability of the lysis circuit to biotechnological applications by employing plasmid DNA extraction as a test bed. We envision that our engineered auto- regulatory circuits that enable programmable functionalities can be readily applied to a range of biotechnological applications.

Functional Probiotics Engineered to Eradicate P. Aeruginosa

Hwang, I. Y., National University of Singapore
Chang, M. W., National University of Singapore

There is a growing sense of urgency in searching for a new effective treatment regimen for infectious diseases as the emergence of multi-drug resistant pathogens continue to pose threats. In our previous study, we engineered *Escherichia coli* to effectively sense and kill a human pathogen, *Pseudomonas aeruginosa,* as a proof-of-concept of a novel microbe-based antimicrobial strategy using the priciples of synthetic biology. In this study, we have chosen a probiotic strain, to develop into therapeutic microbes with clinically relevant functionalities to fight against human pathogen. The implemented functionalities include the ability of the probiotic microbes to swim toward *P. aeruginosa* upon sensing and disrupt biofilm, and exert strong antimicrobial activity against gastrointestinal infections. The latest engineered cells were able to demonstrate pathogen specific cell migration that can sense and kill the pathogen. Furthermore, this target localization was observed in the presence of mature biofilm, which is responsible for antibiotic-resistance and chronic infection, thereby exhibiting significantly improved killing efficiency compared to its non-motile counterpart. The effective antimicrobial activities against target human pathogens have also been shown in eukaryotic infection models. This study suggests the possibility that probiotics could potentially be engineered for prevention and treatment of infectious diseases, which may provide an antimicrobial strategy that is complementary to current antibiotic therapies.

Development and Characterization of Dynamic Controllers in Saccharomyces Cerevisiae
Teo, W. S.,
Hee, K. S.,
Chang, M. W., National University of Singapore

Development and Characterization of Dynamic Controllers in Saccharomyces Cerevisiae

Wei Suong Teo, Kai Sheng Hee, Matthew Wook Chang
Department of Biochemistry, Yong Loo Lin School of Medicine, National University of Singapore, 14 Medical Drive, 117599, Singapore
Synthetic Biology Research Consortium, National University of Singapore, 28 Medical Drive, 117456, Singapore
Expression of heterologous proteins in metabolic engineering endeavours can be detrimental to host cells due to increased usage of cellular resources. Dynamic controls, where protein expression can be triggered on-demand, are effective for the engineering and optimization of bio-catalysts towards robust cell growth and enhanced biochemical productivity. First, we developed and characterized AND-gate dynamic controllers in *Saccharomyces cerevisiae* which combine two dynamic control strategies, inducible promoters and sensing-regulation. These dynamic controllers were constructed based on synthetic hybrid promoters. Promoter enhancer sequences were fused to a synthetic GAL1 core promoter containing DNA binding sites for the binding of a repressor that reduced DNA affinity upon interaction with key intermediates in a biochemical pathway. As fatty acids are key intermediates for production of fatty acid derived advanced biofuels, we used bacterial FadR repressor and operator to demonstrate the functionality of the dynamic controllers. We established that the synthetic GAL1 core promoter can be used as a modular promoter part for constructing synthetic hybrid promoters and conferring fatty acid inducibility. We further showed the performance of the AND-gate dynamic controllers, where two inputs were required to switch the AND-gate ON. Second, we developed genetically encoded xylose sensor-regulators in yeast, where potential applications include dynamic regulation of xylose assimilating pathway enzymes and screening for efficient xylose transporters. Xylose sensor-regulators were created using bacterial XylR repressors and synthetic promoters with XylR operators. The operators are bound by XylR in the absence of xylose to repress transcription. Upon xylose detection, XylR DNA binding activity is antagonized, and the downstream yEGFP reporter expression increases. XylR was shown to repress gene expression from the synthetic promoters and exogenous xylose was able to regulate gene expression. We envision that these synthetic dynamic controllers will function as important tools for engineering and optimizing genetically engineered yeasts towards biochemical

production.

Transporter Engineering for Improved Microbial Tolerance Against Biofuel Molecules

Chen, B., National University of Singapore
Foo, J. L., National University of Singapore
Leong, S. S. J., Nanyang Technological University
Chang, M. W., National University of Singapore

Transporter Engineering for Improved Microbial Tolerance against Biofuel Molecules

Binbin Chen1,2, Jee Loon Foo1,2, Susanna Su Jan Leong1,2,3 and Matthew Wook

Chang1,2

1 Department of Biochemistry, Yong Loo Lin School of Medicine, National University of Singapore, 14 Medical Drive, 117599, Singapore

2 Synthetic Biology Research Consortium, National University of Singapore, 28

Medical Drive, 117456, Singapore

3 Singapore Institute of Technology, 10 Dover Drive, 138683, Singapore

Hydrocarbon alkanes, components of major fossil fuels, are considered as next-generation biofuels because their biological production has recently been shown to be possible. However, high-yield alkane production requires robust host cells that are tolerant against alkanes, which exhibit cytotoxicity. To tackle this toxicity issue, we first investigated the mechanisms of cellular response to alkane biofuels at a system level through transcriptome analyses in *Saccharomyces cerevisiae*, a model eukaryotic host of industrial significance. Among hypothesized mechanisms, we were interested in identifying plasma membrane transporters which could possibly aid in alkane secretion, leading to improved tolerance. In support of this hypothesis, we expressed the hypothesized native and heterologous transporters, and demonstrated that the expression of transporters significantly improved cell tolerance against alkanes through maintaining lower intracellular alkane level. Here, we demonstrated that transporter engineering - identification and expression of transporters led to significantly improved tolerance against alkane biofuels in *S. cerevisiae*. Further, using directed evolution, we isolated AcrB mutants with improved efflux efficiency in *Escherichia coli* for hydrocarbon-based biofuels. The utilization of such optimized native efflux pumps will increase productivity of biofuels synthesis and alleviate toxicity and difficulties in production scale-up in current microbial platforms. We believe that our results laid the groundwork for developing robust microbial cells through transporter engineering, which will greatly aid in next- generation biofuel production and recovery.

Mazf-Mediated Deletion System for Genome Engineering in Budding Yeast Saccharomyces Cerevisiae

Liu, Q., College of Life Sciences, Nankai University
Wu, Y., College of Life Sciences, Nankai University
Zhang, X., College of Life Sciences, Nankai University
Bai, Y., College of Life Sciences, Nankai University
Xu, H., College of Life Sciences, Nankai University
Qiao, M., The Key Laboratory of Molecular Microbiology and Technology, Ministry of Education, Nankai University

mazF-mediated deletion system for genome engineering in budding yeast

Saccharomyces cerevisiae

Background: In the field of functional genomics and metabolic engineering in the postgenome era, the concept of minimum genome factories (MGFs) has been proposed, which can be defined as recombinant strains whose metabolism has been streamlined to the optimal minimal subset in order to maximize product formation for targeted applications (Ara et al., 2007; Giga-Hama et al., 2007; Mizoguchi et al.,

2007). Evolutionary biology imparts us that a certain inherent plasticity exists in the genomes of living organisms and has constituted a major evolution power (Trevors, 1997). The plasticity of genomes gives us hints that it is probable to streamline physiological pathways by means of genome rearrangements, as cells harbor a number of genes that only marginally contribute to cellular fitness and various dispensable activities that are not essential for survival, particularly when cells are grown in rich media. The unicellular eukaryote budding yeast *Saccharomyces* cerevisiae has long been exploited for industrial and biotechnological applications and is extensively chosen as the starting platform for value-added metabolites biosynthesis and biofuel production in the wake of emerging synthetic biology and systems

biology (Ostergaard et al., 2000; Nielsen et al., 2013). Construction of a set of chassis cells with various genetic alterations provides the opportunity to obtain specific target strain(s) with desired performance through phenotype comparisons and genetic regulation analysis and thus represents the fundamental work for the generation of featured MGFs. A chassis member generally contains at least one chromosomal modification, including single gene deletion or removal of a large chromosomal region, and one unmarked chassis without undesirable foreign sequence is preferred. To this end, the employed genetic method should possess two other peculiarities: marker rescue and scarless modification, besides the ability to directly accomplish desired deletion.

Results: In the present report, we proposed a new scarless modification strategy, which evolved from a recently established *mazF*-based gene deletion protocol, for large-scale and scarless genome rearrangements in yeast *S. cerevisiae*. The main experimental process of the presented system is substantially similar to the existing gene deletion methods involving double-crossover integration and single-crossover pop-out except for the way to construct deletion cassette and to accomplish the deletion of large chromosomal targets (Fig. 1). The deletion cassette is composed of five functional parts: a *URA3* marker for integrant selection, a galactose-induced *mazF* cassette for counter-selection and three genome segments. Each part for the deletion cassettes was individually amplified by PCR and co-transformed into yeast cells, and then the complete deletion cassettes were assembled *in vivo* through homologous recombination among the 50 bp sequence overlaps introduced by primers. Homologous recombination among two flanking homology regions and their counterparts in the genome resulted in the targeted integration of deletion cassettes and created new direct repeats of the third interior genome sequence. Integrants were selected on SC-Ura plates and successful integrations were verified by direct colony PCR. The resultant cells were subjected to counter-selection step to accomplish the excision of genomic sequences and recovery of genetic markers resulting from the single cross-over between newly generated direct repeats. This method was demonstrated to be highly efficient for the direct deletion of a designed internal or terminal chromosomal region. Enhanced targeting efficiency of the deletion cassette constructs was achieved through extending the length of flanking homology arms. Furthermore, the developed *mazF*-based counter-selection system outperformed traditional *URA3*/5-FOA method that a greater number of resultant colonies appeared with higher percentage of desirable deletants. Despite the loss of 12 and 10 predicted ORF(s), respectively, upon the deletion of the 26.5-kb and 28.9-kb regions, the strains NK11 and NK12 still exhibited normal growth under standard laboratory conditions; and similar growth profile was also observed for mutant NK13 carrying both these two deletions compared to that of the parental strain.

Conclusion: This *mazF*-based technique can be used to generate scarless and sequential chromosomal modification mutants of *S. cerevisiae* efficiently, using deletion cassettes assembled *in vivo* by means of the recombination machinery of yeast cells. Based on the precision of its deletion, this method may provide a new way to construct genetically modified *S. cerevisiae* strains with large-scale genome deletion more efficiently and advance the regulatory network study of this organism and further genome streamlining.

Keywords: Saccharomyces cerevisiae; mazF; large-scale deletion; counter-selection;

genome engineering

Reference:

Ara, K., Ozaki, K., Nakamura, K., Yamane, K., Sekiguchi, J.,Ogasawara, N., 2007.

Bacillus minimum genome factory: effective utilization of microbial genome information. Biotechnol Appl Biochem 46, 169-178.

Giga-Hama, Y., Tohda, H., Takegawa, K.,Kumagai, H., 2007. *Schizosaccharomyces pombe* minimum genome factory. Biotechnol Appl Biochem 46, 147-155. Mizoguchi, H., Mori, H.,Fujio, T., 2007. *Escherichia coli* minimum genome factory. Biotechnol Appl Biochem 46, 157-167.

Nielsen, J., Larsson, C., van Maris, A.,Pronk, J., 2013. Metabolic engineering of yeast for production of fuels and chemicals. Curr Opin Biotechnol.

Ostergaard, S., Olsson, L.,Nielsen, J., 2000. Metabolic engineering of *Saccharomyces cerevisiae*. Microbiol Mol Biol Rev 64, 34-50.

Trevors, J.T., 1997. Evolution of bacterial genomes. Antonie Van Leeuwenhoek 71, 265-270.

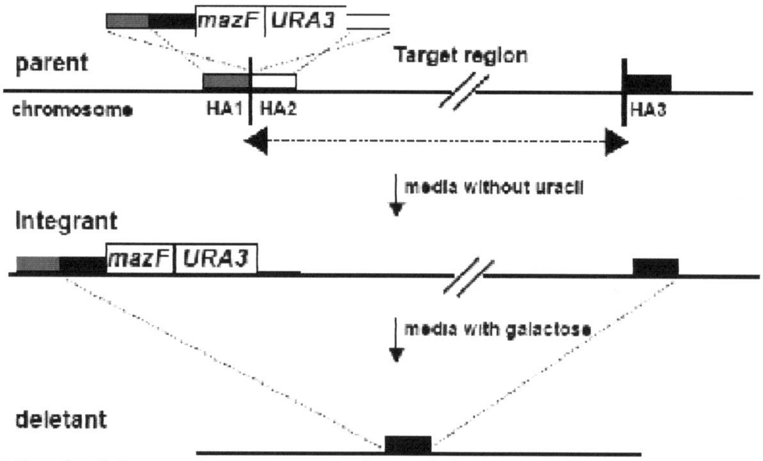

Fig. 1. Schematic representation of the *mazF* mediated scarless deletion of *Saccharomyces cerevisiae* genomic region(s). HA1-3 are short segments of *S. cerevisiae* genome; *URA3* and *mazF* markers are from plasmids pUG72 and pLC1, respectively. These segments were amplified by PCR, co-transformed and formed functional deletion cassette(s) through *in vivo* homologous recombination. The deletion cassette(s) was integrated by double crossover and the deletion of target region along with markers by single crossover.

Creating Components for Synthetic Gene Networks

Shis, D.,
Hussain, F.,
Meinhardt, S.,
Swint-Kruse, L., University of Kansas Medical Center
Bennett, M.,

The ability to implement multi-input gene regulation has been a limiting factor in the construction of synthetic gene networks. This limitation becomes more prescient as researchers seek to go beyond simple synthetic gene networks. In this study we develop methods for multi-input transcriptional activation and repression. For multi-input transcriptional activation, we split T7 RNA polymerase (RNAP) to create a two input transcriptional AND gate whereby gene expression is observed if and only both fragments of the split protein are present. We then modify the specificity of split T7 RNAP to create functionally orthogonal split T7 ORFs. For multi-input transcriptional repression, we co-express multiple lacI/galR chimeric repressors whereby each chimeric repressor regulates an operator site dictated by its DNA binding domain according to inducer sensed by its ligand binding domain. Co-expressing chimeric repressors enmasse creates multi-input regulatory networks such as a four input transcriptional AND gate or even two functionally orthogonal transcriptional AND gates. Altogether our work demonstrates strategies for engineering multi-input transcriptional regulation. These tools will enable the construction of the next generation of synthetic gene networks.

Development of a Synthetic Biology Toolbox for Synechococcus Sp. PCC 7002

Ruffing, A., Sandia National Laboratories

Abstract:

Biofuel and chemical production in cyanobacterial hosts has recently expanded due to the search for renewable alternatives for conventional products of the petroleum industry. Cyanobacteria require minimal resources, utilizing sunlight for energy and carbon dioxide as a carbon source, with some strains even capable of nitrogen fixation. Hence, cyanobacterial-based production offers the promise of reduced feedstock costs and the potential for reduced greenhouse gas emissions compared to traditional fermentation hosts. Despite these potential advantages, cyanobacterial strain development is limited by a

lack of genetic tools and chassis organisms for systems-level optimization. To advanced synthetic biology capabilities for cyanobacteria, this work aims to develop a synthetic biology toolbox (i.e. characterized genetic tools and parts) for the genetic manipulation of *Synechococcus* sp. PCC 7002. Tools developed for *Synechococcus* sp. PCC 7002 include fluorescent protein reporters and native promoters with varying expression levels (strong, moderate, and weak) and regulation (constitutive, linear phase, and stationary phase). Additionally, transformation conditions, including cell concentration, length of homologous regions, and the effect of light and temperature on ploidy level, were investigated for the cyanobacterial chassis *Synechococcus* sp. PCC 7002. The construction and availability of a synthetic biology toolbox is a critical step in advancing cyanobacterial strain development for commercial applications.

***** Formatting Issue - Best Available Paper/Graphic *****

Temporal Inactivation of DNA Repair Enables Highly Precise Genome Engineering in Escherichia coli

Nyerges, A., Biological Research Centre of the Hungarian Academy of Sciences
Csörgo, B., Biological Research Centre of the Hungarian Academy of Sciences
Pósfai, G., Biological Research Centre of the Hungarian Academy of Sciences
Pál, C.,

Temporal inactivation of DNA repair enables highly precise genome

engineering in *Escherichia coli*

Ã□kos Nyerges, BÃ¡lint CsÃ¶rgÅ?, GyÃ¶rgy PÃ³sfai, Csaba PÃ¡l

Synthetic and Systems Biology Unit, Institute of Biochemistry, Biological Research Centre of the Hungarian
Academy of Sciences, Szeged, Hungary http://group.szbk.u-szeged.hu/sysbiol/

The emergence of advanced *in vivo* genome engineering technologies is transforming biotechnology, systems, and evolutionary biology. By exploiting the customized engineering of living organisms, these techniques have redefined the repertoire of experimental protocols. One of the methods, Multiplex Automated Genome Engineering (MAGE), uniquely allows simultaneous targeting of many genomic loci, and thereby enables the generation of huge sequence diversity within a practical timescale and at a reasonable cost.
However, MAGE has one major â??Achillesâ?? heelâ?□, which limits its practical application. As the methyl-directed mismatch repair system acts against the incorporation of desired genomic modifications, MAGE requires mutants with endogenous mismatch repair system deficiency. As a by-product, inactivation of the hostâ??s mismatch repair system results in a dramatically elevated general mutation rate and therefore leads to the accumulation of background mutations across the whole genome. Accumulation of such undesired genetic alterations can dramatically alter the studied effects of the targeted modifications.
Our work presents a novel strategy for mismatch repair evasion using temperature sensitive DNA repair mutants and a method for temporal inactivation of the mismatch repair protein complex, which enables the transient suppression of DNA repair during ssDNA- recombineering in *Escherichia coli*. By restricting the cells mutator state only to the relatively brief oligonucleotide integration period, the number of off-target mutations was reduced by
85%, concurrently maintaining highly efficient and unbiased allelic replacement. This novel
strategy therefore enables a more precise *in vivo* genome-editing approach with reduced off-target effects. This will help future genome engineering endeavors and allow for the investigation of uncharted cellular interactions and several central evolutionary issues.

Citation:

Optimizing Multi-Gene Biological Systems Using High-Throughput DNA Assembly, Sequencing, and Model-Guided Search Strategies

Woodruff, L. B. A., Broad Institute of MIT and Harvard
Mikkelsen, T., Broad Institute of MIT and Harvard
Smanski, M. J., Massachusetts Institute of Technology
Gordon, D. B., Broad Institute of MIT and Harvard
Voigt, C. A., University of California, San Francisco
Nicol, R., Broad Institute of MIT and Harvard

We have developed a framework to modularize and re-optimize massively multi- gene pathways or biological systems that increases the size and complexity of genetic systems that can be engineered. Engineering and optimizing the performance of large multi-gene systems requires combinatorially tuning gene expression through many design iterations. Here, new high-throughput DNA assembly strategies were developed that leverage advances in DNA sequencing to speed this process and also enable statistical combinatorial design for targeted searches through the high-dimensional space of possible genetic designs. We demonstrated these strategies developed here by optimizing a pathway of 16 genes. This 16-gene nitrogen fixation (*nif*) gene cluster is entirely modular and synthetic (â??refactoredâ?□) with ~100 parts. The synthetic *nif* pathway encodes
16 essential genes that have tightly coordinated action to convert atmospheric nitrogen into ammonia, a critical process for global agriculture productivity. Using our novel assembly strategy, we constructed and sequenced a library of >55,000 unique *nif* transcriptional units (TUs). This is a 100-1,000-fold improvement in library size over our and comparable previous studies. Each TU is composed of characterized genetic parts (5-7 each), including one refactored *nif* gene. These *nif* TU modules can be directly assembled into whole 16-gene, 100-part permuted cluster designs (~1045 unique cluster designs possible from this library). Additionally in this study, we sequenced approximately 7.5 million *nif* TUs to characterize DNA assembly error modes and quantify their frequencies. Next, from the library of *nif* modules we built and tested systematically permuted refactored clusters to investigate optimizing nitrogen fixation activity by tuning expression within the refactored cluster using characterized biological parts. Combinatorial design and statistical design of experiments were applied to constrain the part substitutions integrated into the cluster designs fabricated and tested in each round. Together, these methods enable model-guided redesign and optimization of genetic system performance with minimized DNA assembly required (~10-fold reduction in cloning). Looking ahead, we believe the assembly and optimization strategies demonstrated here will accelerate transferring complex functions to engineered systems and dramatically expand the complexity of biological systems that can be written, engineered, and optimized.

Multiplexed and Programmable Regulation of Gene Networks with an Integrated RNA and Crispr/Cas Toolkit in Human Cells

Nissim, L.,
Perli, S. D.,
Fridkin, A.,
Perez-Pinera, P.,
Lu, T. K., Massachusetts Institute of Technology

Multiplexed and Programmable Regulation of Gene Networks with an Integrated RNA and Crispr/Cas Toolkit in Human Cells

Lior Nissim1#, **Samuel D. Perli**2# Alexandra Fridkin1, Pablo Perez-Pinera1 and Timothy K.

Lu3,

(#) These authors contributed equally

(1) MIT, Cambridge, MA,

(2) EECS, MIT, Cambridge, MA,

(3) EECS/Biological Engineering, MIT, Cambridge, MA

CRISPR/Cas transcription factors (CRISPR-TFs) have the potential to enable multiplexed modulation of synthetic and endogenous gene networks in a programmable fashion. However, gRNAs have so far have been expressed only from RNA polymerase III promoters, limiting the integration of CRISPR-TFs with mammalian RNA regulatory architectures. We present multiple strategies for the production of gRNAs in human cells from endogenous, RNA polymerase II promoters by employing Cas6/Csy4 based RNA processing, RNA-triple-helix structures, introns and ribozymes. In addition to generating multiple gRNAs from single transcripts and implementing multi-stage cascades, we integrate CRISPR- TFs with microRNAs to build RNA-dependent networks that can be rewired with Csy4 to achieve complex behaviors. Our toolkit enables the design of scalable gene circuits and the modulation of endogenous regulatory networks for biology, therapeutic, and synthetic biology applications.

Reconstitution of Orphan Fungal Polyketide Gene Clusters in Saccharomyces Cerevisiae

Li, J.,
Harvey, C., Stanford University
Hillenmeyer, M., Stanford University
Davis, R., Stanford Genome Technology Center

Natural products (also known as secondary metabolites) broadly refer to organic compounds formed in living organisms, but are usually taken to mean small molecule metabolites that are not required for growth and development. More than half of FDA-approved antibiotics and anticancer drugs are natural products (NPs) or their derivatives. In particular, fungi produce a wide range of bioactive compounds such as penicillin, whose discovery kicked off modern era of antibiotic discovery, and lovastatin, a cholesterol-control drug that generated billions of dollars in annual sales.
The landscape of fungal natural products has been changing since the explosion of genomic data. Many NP gene clusters can now be readily identified based on homology, but no associated NP could be identified as being produced by the organism, and often the genes were not even expressed. Those biosynthetic pathways whose secondary metabolites are unknown are defined as â??orphanâ?□ gene clusters. The thousands of orphan fungal gene clusters computationally identified may encode a wealth of unknown natural products that may exhibit attractive bioactivities.
To explore various NP clusters, researchers have heterologously expressed them in various model organisms, among which S. cerevisiae has emerged as a powerful host. Our project aims to explore some of the thousands of orphan fungal gene clusters using *S. cerevisiae* as a host to heterologous express the fungal gene clusters. We hypothesize that more natural product clusters can be explored by bypassing expression in native hosts, and instead expressing the predicted gene clusters in the heterologous host, *Saccharomyces cerevisiae*. We explore two orphan clusters: one in *Aspergillus fumigatus*, a common pathogenic fungal strain that causes disease in immune-comprised patients. The other is in *Trichoderma virens*, an opportunistic avirulent fungal species that protect many plants from various pathogens.
We designed and purchased codon-optimized synthetic DNA fragments and will assemble them into our yeast expression system. Once the genetic tools are constructed, we will access protein expression via western blotting and use LC/MS to determine the molecular formula of products from the pathway. NMR structural characterization will be performed subsequently. We will also determine the structure of secondary metabolic intermediates and elucidate the order of reactions to understand the function and specificity of the enzymes involved.
Our work could provide an alternative traditional natural product chemistry to access natural product and a platform for detailed enzymatic and biochemical studies on the function and specificity of enzymes involved in natural product biosynthesis.

Engineering Regulation in Anaerobic Fungi During Lignocellulose Breakdown

John Henske, Kevin Solomon, Sarah Maxel, Michelle Oâ??Malley

A significant struggle in the production of biofuels and other value-added products from agricultural waste is the breakdown of complex, lignin-rich material in a cost efficient manner. Our lab has targeted fungi, isolated from the anaerobic microbiome of large herbivores, as a source for a variety of unique enzymes that work together to break down this complex material. My primary interest is to investigate how the fungi regulate these enzymes, and to manipulate these processes to develop gut fungi as novel biofuel production platforms. We have investigated transcriptional regulation through the use of next generation RNAseq to determine how the expression levels of cellulose-degrading enzymes change in an isolated strain of *Piromyces* fungus. Initial experimental results suggest that the production of biomass-degrading enzymes is regulated through catabolite repression in the presence of simple monosaccharides, which suggests that proteins of unknown function can be identified by mapping regulation patterns in response to substrate. Additional experiments where the fungi are grown in the presence of both crystalline cellulose (filter paper) and monosaccharides show a decrease in utilization of cellulose material when simple sugars are present, which supports the hypothesized regulation mechanism. Surprisingly, this repression of cellulose-degrading behavior exists even in the presence of monosaccharides that do not support fungal growth. In light of these results, we hypothesize that non- specific sugar sensing mechanisms underlie functional reprogramming of cellulase secretion. The aims of my project are to (1) identify and verify receptor-mediated sensing mechanisms through RNAseq and biochemical studies; (2) apply differential expression analysis to identify novel proteins that are regulated by these sensing mechanisms; and (3) determine the flux of carbon through anaerobic fungi to reveal strategic targets for altering metabolism and substrate utilization. These efforts are underway for our first isolated strain (*Piromyces*) and are being expanded to two additional strains, representing the *Neocallimastix* and *Anaeromyces* genera of anaerobic fungi, as additional sources for regulatory machinery. With an understanding of these regulation mechanisms, they may be used to establish a more efficient process for bio-based production from agricultural that tunes enzyme production based on the most abundant nutrient sources available.

Design and Implementation of a Biomolecular Concentration Tracker

Hsiao, V., California Institute of Technology
de los Santos, E. L. C., California Institute of Technology
Whitaker, W. R., Stanford University
Dueber, J. E., U.C. Berkeley
Murray, R. M., California Institute of Technology

As a field, synthetic biology strives to engineer increasingly complex artificial systems in living cells. Active feedback in closed loop systems offers a dynamic and adaptive way to ensure constant relative activity independent of intrinsic and extrinsic noise. In this work, we use synthetic protein scaffolds as a modular and tunable mechanism for concentration tracking through negative feedback. Input to the circuit initiates scaffold production, leading to colocalization of a two-component system and resulting in the production of an inhibitory anti-scaffold protein. Using a combination of modeling and experimental work, we show that the biomolecular concentration tracker circuit achieves dynamic protein concentration tracking in *Escherichia coli* and that steady state outputs can be tuned.

Rapidly Characterizing the Fast Dynamics of RNA Genetic Circuitry with Cell-Free Transcription-Translation (TX-TL) Systems

Takahashi, M. K., Cornell University
Chappell, J., Cornell University
Hayes, C. A., California Institute of Technology
Sun, Z. Z., California Institute of Technology
Kim, J., California Institute of Technology
Singhal, V., California Institute of Technology
Spring, K. J., University of Texas Medical School at Houston
Al-Khabouri, S., Universitex de Montreal
Fall, C. P., University of Illinois
Noireaux, V., University of Minnesota
Murray, R. M., California Institute of Technology
Lucks, J. B., Cornell University

The ability to engineer synthetic genetic networks using RNA regulators has solidified them as powerful components of the synthetic biology toolbox. A potential strength of RNA regulators is that they have fast degradation rates compared to proteins, and may be able to propagate signals through networks on the timescale of RNA degradation. However, a current bottleneck to understanding signal propagation through genetic networks is the slow design-build-test cycle of evaluating them *in vivo*. In this work we adapt an *Escherichia coli*-based cell-free transcription-translation (TX-TL) system for rapidly prototyping RNA transcriptional networks. The flexibility of the TX-TL system allowed us to design an experiment to measure the response time of a two-stage RNA transcription cascade, which we found to be approximately 10 minutes, or about 5 minutes per step of the cascade. We then show that this response time can be adjusted by regulator threshold tuning, which we achieved using tandem RNA regulators in front of target genes. This allowed us to prototype a new RNA network, an RNA single input module (RNA SIM), which was predicted to sequentially activate the expression of two different genes. Finally, we show that the RNA SIM functions as designed *in vivo*. This work lays the foundation for using TX-TL as a tool for RNA circuit prototyping. We conclude with thoughts on what needs to be done to quantitatively translate TX-TL circuit characterization into predictions for *in vivo* circuit function.

Isocost Lines in the Cellular Economy: Gene Expression Is Coupled Due to Competition for Shared Resources

Gyorgy, A., MIT
Jimenez, J. I.,
Chung, H.,
Weiss, R., M.I.T.
Del Vecchio, D.,

Isocost lines in the cellular economy: gene expression is coupled due to competition for shared resources

Andras Gyorgy1,*, Jose I. Jimenez2,*, Hattie Chung3,
Ron Weiss1 and Domitilla Del Vecchio1

1Massachusetts Institute of Technology, Cambridge USA

2University of Surrey, Guildford UK

3Harvard Medical School, Cambridge USA

*equal contribution

Transcriptional and translational resources are available in limited amounts. Since genetic circuits in living cells share these common resources, unexpected couplings can arise among seemingly unconnected genetic modules, yielding poorly predictable circuit behavior. Among the cellular resources required for gene expression, RNAP and ribosomes are key factors. Therefore, characterizing how gene expression becomes coupled due to the limited availability of RNAP and ribosomes is essential both for understanding the behavior of natural circuits and for engineering new ones.
Using a physics-based model together with experiments on synthetic constructs in *E. coli*, we discover that the expression of two genes without a regulatory link between them is linearly constrained. This linear relationship can be interpreted as an isocost line, introduced in microeconomics to describe what combinations of two products can be purchased with a limited budget. We characterize the extent by which proteins become coupled due to the limited availability of RNAP and ribosomes: it is substantial ranging from 60% when genes are on medium copy number plasmids to
30% when genes are on low copy number plasmids. These effects are therefore significant in the design of synthetic gene circuits even when assembled on low copy number plasmids.
Furthermore, as RBS strength and plasmid copy number are two of the most commonly tuned experimental parameters, we characterize how changing them affects salient

features of the isocost line, such as its slope, establishing the extent of coupling. To this end, we construct a system with two fluorescent reporter proteins, one inducible and one constitutive without a regulatory path between them on the same plasmid, for several combinations of plasmid copy number and RBS strength for the inducible gene. In agreement with the mathematical predictions, changing the RBS strength of the inducible gene rotates the isocost line, while changing the plasmid copy number shifts the isocost line.

Our work demonstrates that isocost lines can be employed to reliably predict how the behavior of genetic circuits become coupled when sharing limited cellular resources, and that they provide design guidelines for minimizing the effects of such couplings. For instance, our results reveal, perhaps counter-intuitively, that choosing the strongest RBS for the inducible gene is the best strategy to minimize its effect on the expression of the constitutive gene.

Engineering of Small RNA Transcriptional Activators

Chappell, J., Cornell University
Takahashi, M. K., Cornell University
Lucks, J. B., Cornell University

RNA regulators of gene expression have become an integral component of the synthetic biology toolbox. Small RNAs (sRNAs) are an abundant and mechanistically diverse class of RNA regulator that can affect multiple aspects of gene expression. However, there are no known natural or synthetic examples of sRNAs that can directly activate transcription *in vivo.* Motivated by this natural gap in RNA based regulation, we re-engineered the structural logic of a naturally occurring RNA transcriptional attenuator to create sRNA transcriptional activators that function in *Escherichia coli*. Using two different mechanistic designs, we show that we can create *trans*-acting sRNAs that promote the disruption of RNA intrinsic transcriptional terminators located upstream of a target gene. Through a series of optimizations, we were able to achieve up to 90-fold activation of transcription. Moreover, by applying this strategy to a different RNA transcriptional attenuator, and to a transcriptionally repressive riboswitch, we showed that we can create additional transcriptional activators that are both orthogonal to each other, and to existing RNA transcriptional repressors. Our results demonstrate that sRNAs are capable of activating transcription, and provide a new mechanism for the regulation of gene expression.

Targeting Chromatin: An Emerging Layer of Synthetic Control

Keung, A. J., Howard Hughes Medical Institute
Bashor, C. J., Howard Hughes Medical Institute, Boston University
Kiriakov, S., Boston University
Collins, J. J., Boston University
Khalil, A. S., Boston University

The complex structure of chromatin vastly extends the information potential of the eukaryotic genome beyond the underlying DNA sequence. Several hundred chromatin regulators (CRs) modulate eukaryotic genome structure and function by remodeling nucleosomes and DNA. Yet, despite being the subject of extensive studies, it remains unclear how CRs operate through diverse processes to regulate gene transcription, and this powerful layer of gene regulation has yet to be exploited in engineering applications. Synthetic biology offers a unique approach to decompose and exploit the complexity of chromatin by studying emergent transcriptional behaviors from engineered combinatorial, spatial, and temporal patterns of individual CRs. Here we fuse a library of 223 yeast CRs to programmable zinc finger proteins. Site-specific and combinatorial recruitment of CRs to engineered genomic loci drives a host of transcriptional logic functions, spatial behaviors, and epigenetic control. Analyzing these transcriptional behaviors through gene ontology provides useful design principles for CR organization in establishing and engineering chromatin-based transcriptional regulation. This work presents a bottom-up approach to investigating chromatin regulation, and introduces new, chromatin-based components, useful transcriptional behaviors, and design strategies for synthetic biology and cellular engineering.

Gene Circuit Performance Characterization and Resource Usage in a Cell-Free "Breadboard"

Siegal-Gaskins, D., California Institute of Technology
Tuza, Z. A.,
Kim, J.,
Noireaux, V., University of Minnesota
Murray, R. M., California Institute of Technology

The many successes of synthetic biology have come in a manner largely different from those in other engineering disciplines; in particular, without well-characterized and simplified prototyping environments to play a role analogous to wind-tunnels in aerodynamics and breadboards in electrical engineering. However, as the complexity of synthetic circuits increases, the benefitsâ??in cost savings and design cycle timeâ??of a more traditional engineering approach can be significant. We have recently developed an *in vitro* â??breadboardâ? prototyping platform based on *E. coli* cell extract that allows biocircuits to operate in an environment considerably simpler than, but functionally similar to, *in vivo*. The simplicity of this system makes it a promising tool for rapid biocircuit design and testing, as well as for probing fundamental aspects of gene circuit operation normally masked by cellular complexity. In this work, we characterize the cell-free breadboard using real-time and simultaneous measurements of transcriptional and translational activities of a small set of reporter genes and a
transcriptional activation cascade. We determine the effects of promoter strength, gene concentration, and nucleoside triphosphate concentration on biocircuit properties, and we isolate
the specific contributions of essential biomolecular resourcesâ??core RNA polymerase and ribosomesâ??to overall performance. Importantly, we show how limits on resources, particularly those
involved in translation, are manifested as reduced expression in the presence of orthogonal genes that serve as additional loads on the system.

Optimized in-Fusion Cloning System Demonstrates High Efficiency and Accuracy of Multi-Fragment Cloning

Oh, S., Clontech Laboratories, Inc
Jokhadze, G.,
Duong, T.,
Bostick, M.,
Farmer, A.,

Optimized In-Fusion cloning system demonstrates high efficiency and accuracy of multi-fragment cloning

Steve Oh1, Gia Jokhadze1, Tommy Duong1, Magnolia Bostick1, Andrew A. Farmer1

1Clontech Laboratories, Inc., Mountain View, CA

In-Fusion PCR Cloning technology enables directional, seamless cloning of any PCR fragment into any linearized vector in a single 15-minute reaction. No additional treatment of the PCR fragment is required (such as restriction digestion, ligation, phosphorylation, or blunt-end polishing). In contrast, the In-Fusion enzyme generates short regions of single-strand overlap between vector and insert, so as to drive directional cloning of the desired fragment. This overlap can be designed into the PCR primers which amplify the desired sequences. While cloning single inserts is extremely efficient, cloning experiments are becoming increasingly complex, involving multiple fragments to stitch genes together from synthetic building blocks or to engineer new functionalities through combinatorial protein domain swapping, etc. This, in some cases, may present a challenging task. Here, we present data showing multi-fragment cloning efficiency and accuracy can be significantly increased by optimizing various parameters of the In-Fusion Cloning reaction. This extends the versatility of the In-Fusion cloning system to meet the coming challenges of innovative research in the synthetic biology field.

A Low Cost, Customizable Turbidostat for Use in Synthetic Circuit Characterization

Takahashi, C.,
Miller, A.,
Ekness, F.,
Klavins, E.,
Dunham, M.,

Engineered biological circuits are often disturbed by a variety of environmental factors. In batch culture, where the majority of synthetic circuit characterization occurs, environmental conditions vary as the culture matures. Turbidostats are powerful characterization tools that provide static culture environments, however they are often expensive especially when purchased in custom configurations, and are difficult to design and construct in a lab. Here, we present a low cost, open source multiplexed turbidostat that can be manufactured and used with minimal experience in electrical or software engineering. We demonstrate the utility of this system to profile synthetic circuit behavior in \yeast. We also demonstrate the flexibility of the design by showing that a fluorometer can be easily integrated.

A Rapid, Label-Free, and Scalable Method for Characterizing Binding Properties of Small Molecule Aptamers

McKeague, M.,
Chang, A. L., Stanford University
Smolke, C. D., Stanford University

Small molecule-binding aptamers have emerged as versatile sensing and targeting elements for diagnostic, therapeutic, and gene-regulatory applications. Approximately 60 small molecule-binding RNA aptamers have been selected in vitro; however, limited characterization and functional validation have hindered their broader use in synthetic biology applications. For example, the majority of in vivo engineered RNA-based gene regulatory devices exclusively incorporate the theophylline aptamer as the sensing component. One of the challenges associated with developing and implementing new small molecule-binding aptamers is the ability to rapidly and reliably validate their binding properties. Here we describe a rapid and sensitive method for characterizing kinetic and binding affinity properties of small molecule aptamers based on the Biacore surface plasmon resonance biosensor platform. We validated the SPR-based aptamer characterization strategy by comparing the SPR-measured binding properties of previously studied aptamers and further characterized a diverse panel of 12 small molecule-binding RNA and DNA aptamers. This panel included both natural and in vitro selected aptamers, providing insight regarding differences in their binding properties. Our system is label-free, scalable, and supports the analysis of different aptamer–target pairs and binding conditions with the same platform. Thus, the described SPR-based aptamer characterization method provides a versatile and powerful tool to facilitate the rapid validation of RNA aptamer sensors, supporting their broader application in engineering RNA devices and multicomponent systems.

Recombinant Optimization of an Orthogonal Tryptophanyl Suppressor tRNA
Pothukuchy, A., University of Texas at Austin
Hughes, R., University of Texas at Austin

Recombinant Optimization of an Orthogonal Tryptophanyl Suppressor tRNA

Pothukuchy A, Ellefson JW, Ellington AD, Hughes RA*
The engineering of novel synthetic systems and components will lead to the creation of novel biologically derived or inspired products ranging from commodity chemicals to synthetic proteins. For the creation of synthetically augmented proteins it is possible to further expand the diversity of the genetic code beyond 20 amino acids with additional synthetic amino acids incorporation at specific sites in proteins using orthogonal aminoacyl-tRNA synthetase/tRNA pairs. We have previously shown that the tryptophanyl tRNA synthetase/tRNA pair is an attractive candidate for achieving the goal of unnatural amino acid incorporation because of its spacious binding pocket and its potential to accommodate large and bulky synthetic amino acids. Here we report the development of the next generation of orthogonal tryptophanyl-tRNAs that are not only superior in their suppression ability but are also truly orthogonal in *E.coli* system. Rational design was used to create novel tRNA variants by making modifications on the acceptor arm as well as on the anticodon stem region of the parental tRNA sequences. These next generation tRNAs are not only useful for improved incorporation of unnatural amino acids into proteins but are also useful for understanding the capacity of novel orthogonal translation system components for applications in synthetic biology.

References:

1. Ellefson, J.W., Meyer, A.J., Hughes, R.A., Cannon, J. R., Brodbelt, J.S. and Ellington, A.D. (2013). Directed evolution of genetic parts and circuits by compartmentalized partnered replication. Nat Biotechnol., **32**(1): 97-101.
2. Hughes, R..A. and Ellington A.D. (2010).Rational design of an orthogonal tryptophanyl nonsense suppressor tRNA. Nucleic Acids Res., **38**(19): 6813-30.
3. Chatterjee, A., Xiao, H., Yang, P.Y., Soundararajan, G. and Schultz, P.G. (2013).A tryptophanyl-tRNA synthetase/tRNA pair for unnatural amino acid mutagenesis in *E. coli*. Angew Chem Int Ed Engl., **52**(19): 5106-9.

RNA-Guided Nucleases As Programmable-Spectrum Antimicrobials and Microbial Population Sculptors

Citorik, R., Massachusetts Institute of Technology
Mimee, M., Massachusetts Institute of Technology
Lu, T. K., Massachusetts Institute of Technology

With the rising tide of antibiotic resistance and concomitant dearth of novel antimicrobials, new paradigms to treat bacterial infections are in dire need. Unlike conventional, broad-spectrum antibiotics, which target pathogenic and commensal bacterial populations alike, we developed a programmable-spectrum antimicrobial whose activity is dependent on the presence of genetic signatures in microbial cells. By rewiring the Type II CRISPR-Cas system of *Streptococcus pyogenes*, we created RNA-guided nucleases (RGNs) that mediate double-stranded breaks in genes or single-nucleotide polymorphisms (SNPs) associated with drug resistance or virulence. Using bacteriophage particles (ΦRGN) or conjugative vectors (mRGN), RGNs could be efficiently delivered to bacterial populations to enact sequence-specific killing. When targeting a chromosomal SNP which confers resistance to quinolone class antibiotics, ΦRGNs were able to elicit a greater than 4-log reduction in viable cells, while having no activity against the wild-type parental strain. In addition to targeting the chromosome, we found that using ΦRGNs to target the $bla_{\mathrm{NDM-1}}$ and $bla_{\mathrm{SHV-18}}$ genes on multidrug resistance plasmids caused a 2-3-log reduction through the artificial activation of co-harbored toxin-antitoxin systems. Overexpression of the cognate antitoxin resulted in curing of pSHV-18 and antibiotic resensitization. To further demonstrate specificity of our particles, we treated a consortium of three *E. coli* strains with ΦRGNs against the unique resistance markers of each strain and observed selective killing of only the desired strain. Finally, we designed constructs against a virulence gene in enterohemorrhagic *E. coli* O157:H7 and used the ΦRGNs to significantly improve survival in a wax moth larva infection model.

Conditional siRNA Production in Human Cell Lysate Via Shape and Sequence Transduction with Small Conditional RNAs

Hochrein, L. M., California Institute of Technology
Ge, T. J., California Institute of Technology
Schwarzkopf, M., California Institute of Technology
Pierce, N. A., California Institute of Technology

RNA interference (RNAi) mediated by small interfering RNAs (siRNAs) enables knockdown of a gene of choice, executing the logical operation: silence gene Y. The fact that the siRNA is constitutively active is a significant limitation, making it difficult to confine knockdown to a specific locus and time. To achieve spatiotemporal control over silencing, we seek to engineer small conditional RNAs (scRNAs) that mediate □conditional RNAi□ corresponding to the logical operation: if gene X is transcribed, silence independent gene Y. By appropriately selecting gene X, knockdown of gene Y could then be restricted in a tissue- and time-specific manner. To implement the logic of conditional RNAi, we recently engineered scRNAs that, upon binding to mRNA □detection target□ X in a test tube, perform shape and sequence transduction to form a Dicer substrate targeting independent mRNA □silencing target□ Y, with subsequent Dicer processing yielding an siRNA targeting mRNA Y for destruction (Hochrein *et al.*, *J Am Chem Soc, 2013*). Here, we demonstrate conditional siRNA production in human cell lysate; signal transduction is carried out by a single scRNA in a single step. These results demonstrate that scRNA-mediated shape and sequence transduction can proceed in the presence of cellular proteins and nucleic acids and that Dicer cleaves only the signal transduction product, achieving an important step towards the goal of establishing a new paradigm for programmable conditional regulation based on signal transduction with small conditional RNAs.

Synthetic Biology with a Cell-Free TX-TL System: Metabolism, Gene Circuits, Phages and Artificial Cell

Shin, J., MIT
Caschera, F., University of Minnesota
Noireaux, V., University of Minnesota

Cell-free transcription-translation (TX-TL) systems are becoming valuable platform for synthetic biology. *In vitro* TX-TL is now used for many applications ranging from protein evolution to therapeutics and high-throughput production. Noireaux's lab has been interested in using cell-free TX-TL to execute DNA programs *in vitro*.

I will present a unique cell-free TX-TL platform specifically engineered to construct biochemical systems in test tubes (Figure 1). Our research project has three parts. We develop new **metabolism** pathways to fuel cell-free TX-TL reactions. We showed that simple carbon sources are as efficient as the expensive phosphate donor-kinase systems to regenerate ATP. Not only protein synthesis is strong, the degradation of the synthesized proteins using the endogenous AAA+ proteases is also much more efficient. We construct **synthetic gene circuits**, such as regulatory networks, to make quantitative models of gene regulation. The goal is to use these circuit modules to coordinate the expression of complex cellular functions. Recently we demonstrated that our platform can execute large DNA programs that encodes complex **self-assembled entities**. The phage T7 composed of about 60 genes is entirely expressed and synthesized in the cell-free TX-TL reactions. The E. coli MreB cytoskeleton was reconstituted in liposomes. The ultimate goal of this research program is the **synthesis of a minimal cell**. This is done by integrating metabolism, information and self-assembly processes and unraveling the cooperative link between these three indispensable components. I will present the last experiments on those topics.

Spider Dragline Silk Inspired Man-Made Fibers with Unique Property

Peng, C., Clemson University
Russo, J., Clemson University
Gravgaard, C.,
McCartney, H.,
Gaines, W.,
Marcotte, W. Jr., Clemson University

From artificial tissue scaffold and bones to the crafts in aviation and automotive, spider dragline silks are capable to spin their way in. Man-made fibers that resemble the structure and function of spider dragline silk and consistently produced in large quantities are demanding and will bring excitements in medical and engineering applications. We constructed the building blocks of native dragline silk (Spidroin 1, Spidroin 2) from *Nephila clavipes* and inserted into expression plasmids (pKM12 and pKLP36) and transfected tobacco *Nicotiana tabacum*. The plant derived mini-spidroins resembles native N-terminus and C-terminus and 8, 16, or 32 copies of consensus repeat domain of Spidroin1(Sp1) and Spidroin2(Sp2). We confirmed gene insertion and RNA transcription by PCR and reverse-transcriptase PCR, respectively. Mini-sdpridroins are purified through chitin binding domain, an affinity tag that binds onto chitin resin and self-cleavable after chemical activation.

Freeze-dried mini-spidroins formed viscous gelatin-like fluids which exhibited extending continuum property. Immuno-detection of this fluid showed various degree of polymerization of mini-spidroin monomers. When treated with acid, cross-linked by glutaraldehyde and diluted in phosphate buffer (pH 7), the condensed mini-spidroin polyelectrolytes caused negatively charged counterion (gellan gum 0.5%) condensation, forming a thin proteinaceous film at the interface. The film can be pulled into fibers that resistant to water, methanol or ethanol. Single or a combination of various mini-spidroins resulted in various types of fibers. When observed under the microscope, the fibers are auto-florescent and "twisty" with micro-fibril ridges along the fiber axis. Tensile strength testing showed unique physical properties of these fibers.

Enhanced Isoprenoid Production through the Cofactor Metabolic Engineering in Saccharomyces Cerevisiae

Zhu, L., National University of Singapore
Chong, H.,
Ching, C. B., Singapore Institute of Technology

Enhanced isoprenoid production through the cofactor metabolic engineering in

Saccharomyces cerevisiae

Linqi Zhu, **Huiqing Chong**, Chi Bun Ching
Department of Chemical and Biomolecular Engineering, National University of Singapore, Singapore
117585

Abstract:

NADPH, a reducing equivalent, usually plays an important role in the biosynthetic reactions and the anabolic pathways in various microorganisms. The isoprenoid pathway in *Saccharomyces cerevisiae* is largely dependent on the availability of NADPH pools, and therefore increment in NADPH availability is likely to enhance the isoprenoid production.

In this study, we attempt to increase the availability of NADPH to the isoprenoid pathway in *S. cerevisiae* through integration of NADP+-dependent glyceraldehyde 3-phosphate dehydrogenase (*gapC*) from *Clostridium acetobutylicum*. To further enhance the NADPH availability, *gapC* is overexpressed together with (i) native NADH kinase directed to the cytosol (truncated c*POS5*) which mediate the ATP- driven conversion from NADH to NADPH, (ii) pyridine nucleotide transhydrogenase (*udhA*) from *Klebsiella pneumoniae* for the reversible transfer of reducing equivalents between NADH and NADPH, or (iii) glucose 6-phosphate dehydrogenase (*ZWF1*) to increase NADPH regeneration from pentose phosphate pathway The effect of increasing NADPH availability on the isoprenoid pathway was investigated through production of amorpha-4,11-diene.

It was found that integration of *gapC* into the genome of *S. cerevisiae* alone resulted in a 11.9 % increase in amorpha-4,11-diene yield, while overexpression of *gapC* and cPOS5 further enhanced the yield to 5.78 mg/l, an 20.9% improvement as compared to the wild-type (4.78 mg/l). However, when *gapC* was overexpressed with either *udhA* or ZWF1, the amorpha-4,11-diene yield was found to be even lower than the wild-type.

The host characteristics and product yield changes were closely linked to the variation of the intracellular concentration of the cofactors. These results will provide valuable suggestions for the cofactor regeneration and the design of the metabolic networks to improve the isoprenoid production in yeast.

Rapid Evaluation of Tyrosine Kinase Activity of Membrane-Integrated Human Epidermal Growth Factor Receptor Using the Yeast Ggamma Recruitment System

Fukuda, N.,
Honda, S.,

Epidermal growth factor receptor (EGFR) is a member of the receptor tyrosine kinase family and plays key roles in the regulation of fundamental cellular processes, including cell proliferation, migration, differentiation, and survival. Deregulation of EGFR tyrosine kinase activity is involved in the development and progression of human cancers. In the present study, we attempted to develop a method to evaluate the tyrosine kinase activity of human EGFR using the yeast Ggamma recruitment system. Autophosphorylation of tyrosine residues on the cytoplasmic tail of EGFR induces recruitment of Grb2-fused Ggamma subunits to the inner leaflet of the plasma membrane in yeast cells, which leads to G-protein signal transduction and activation of downstream signaling events, including mating and diploid cell growth. We demonstrate that our system is applicable for the evaluation of tyrosine kinase inhibitors, which are regarded as promising drug candidates to prevent the growth of tumor cells. This approach provides a rapid and easy-to-use tool to select EGFR-targeting tyrosine kinase inhibitors that are able to permeate eukaryotic membranes and function in intracellular environments.

A Modular Approach to the Design of Embedded Controllers for Chemical Reaction Networks

Cosentino, C., University of Magna Graecia
Bilotta, M., Universita' degli Studi Magna Graecia di Catanzaro
Montefusco, F.,
Sawlekar, R.,
Amato, F., University of Magna Graecia
Bates, D., University of Warwick

The definition of a general framework for the control of biomolecular circuits represents a challenging objective in the field of Synthetic Biology. While the methods of Control Theory have long since fruitfully contamined many other fields of science and engineering, like aerospace, communication networks or nanotechnology, their application in Synthetic Biology is mainly hampered by the lack of a modular approach to the realization of embedded biological controllers. In this paper we tackle this problem, devising few basic modules that can be assembled to construct a simple closed-loop feedback control scheme. These modules are composed of a minimal set of simple reactions and their description is given at an abstract level, without focusing on a specific biological implementation. To this aim, we have adopted the formalism of Chemical Reaction Networks (CRNs) with mass-action kinetics. A simulated experiment is presented to show how our approach can be used, at least in an ideal setting, to effectively regulate the output of an in-silico CRN process to a desired value. These results, though preliminary, pave the way to new strategies for the design of synthetic biological systems and, more generally, to the establishment of a general control theory for synthetic biomolecular circuits.

Control Engineering Inspired Design Tools for Synthetic Biology

Arpino, J., Imperial College London
Hancock, E., Oxford University
Tomazou, M. III, Imperial College London
Yuan, Y., Cambridge University
Beguerisse, M.,
Goncalves, J., Cambridge University
Barahona, M., Imperial College London
Polizzi, K. M., Imperial College London
Stan, G. B., Imperial College London
Papachristodoulou, A., Oxford University

The development of proof-of-concept experimental applications has revealed that, when constructed in the laboratory, most designs do not work as predicted and need to be fine-tuned on a case-by-case basis. Here we use engineering principles to establish a redesign framework to reliably generate synthetic genetic systems that behave in a predictable fashion. The framework is based on a systems-engineering design cycle. We illustrate the application of this cycle on the toggle switch and other networks, for which we investigate their quantifiably robust and predictable implementation. We first produce detailed mathematical model for the system that captures its most important biochemical properties. Combining prior knowledge reported in the literature with experimental data, we show how the constructed models and their parameters can be systematically identified. Based on a robust control analysis of the constructed models and taking into account model uncertainties, we propose modifications to biological parts that can be "easily□" implemented experimentally so as to achieve an *a priori* specified (re-) design objective. We show how experimental data from the implemented prototype systems can be used to refine the models and improve the design through successive iterations. As an example we demonstrate a model-directed redesign of a toggle switch that meets performance specifications and predictively transits between bistable and monostable behaviours through modifications of ribosome binding site (RBS) strengths.

Controlled Gene Amplification Enables High, Stable, Selection Free Gene Expression in Saccharomyces Cerevisiae

Nødvig, C. S., Technical University of Denmark
Buron, L. D., Technical University of Denmark
Strucko, T., Technical University of Denmark
Jarczynska, Z.,
Mølgård, L., Technical University of Denmark
Mortensen, U. H., Technical University of Denmark

Yeast cell factories can be used to produce a wide range of products, from drugs and vaccines, biofuels, chemical building blocks etc. A widely used strategy to increase the production of a compound of choice in a yeast cell factory is by manipulating the gene copy number. A common way to obtain high gene copy numbers is by the expression from the yeast 2 micron (2μ) multi copy plasmid. The 2μ plasmid enables high expression levels, but comes with inherent genetic instability, since plasmid copies are lost over time, even under selective pressure, and without selection pressure, which is common for industrially relevant media, the plasmid loss is only exacerbated. Furthermore while 2μ often yield high average expression levels, the variance between individual cells in the populations are differs in several orders of magnitude. Genomic integration provides an attractive alternative to plasmid based expression, but brings other concerns, for instance not all chromosomal regions might be able to support high expression due silencing effects, while another concern might be that multiple integrations in the form of tandem repeats risk excision by direct repeat recombination.

To solve some of these issues we have developed a novel gene amplification system, which allows for the construction of stable yeast strains with up to ten integrated gene copies. The gene copy number can be precisely controlled and integration happens at predefined locations, capable of supporting high levels of expression. It is a simple procedure to use the system, only requiring transformation of the genes to be amplified followed by induction and transfer to selective media, which can be done in a timeframe of 15 days. Gene copy number is decided in advance, by choosing one of several strains we have created. Using CFP or β-galactosidase as model proteins a linear correlation between gene copy number and expression levels could be observed, when testing copy numbers between one and seven. Furthermore expression levels of a seven copy strain with CFP equaled that of a strain expressing CFP from a 2μ plasmid. To test the stability and consistency of the system, FACS (Fluorescence-activated cell sorting) experiments were setup, measuring the fluorescence on a single cell level. Results showed that CFP and RFP when integrated in our system, were expressed far more consistently when grown without selection, compared to when RFP and CFP were expressed from a 2μ plasmid and grown under selection pressure. Furthermore the system has successfully been applied to increase the production of both vanillin-glucoside and 6-Methyl Salicylic Acid. Finally it is possible to amplify large pieces of DNA, giving the opportunity to amplify whole pathways, greatly expanding the applications of the system.

Tunable and Multifunctional Eukaryotic Transcription Factors Based on Crispr/Cas

Farzadfard, F.,

Transcriptional regulation is central to the complex behavior of natural biological systems and synthetic gene circuits. Platforms for the scalable, tunable, and simple modulation of transcription would enable new abilities to study natural systems and implement artificial capabilities in living cells. Previous approaches to synthetic transcriptional regulation have relied on engineering DNA-binding proteins, which necessitate multistep processes for construction and optimization of function. Here, we show that the CRISPR/Cas system of *Streptococcus pyogenes* can be programmed to direct both activation and repression to natural and artificial eukaryotic promoters through the simple engineering of guide RNAs with base-pairing complementarity to target DNA sites. We demonstrate that the activity of CRISPR-based transcription factors (crisprTFs) can be tuned by directing multiple crisprTFs to different positions in natural promoters and by arraying multiple crisprTF-binding sites in the context of synthetic promoters in yeast and human cells. Furthermore, externally controllable regulatory modules can be engineered by layering gRNAs with small molecule-responsive proteins. Additionally, single nucleotide substitutions within promoters are sufficient to render them orthogonal with respect to the same gRNA-guided crisprTF. We envision that CRISPR-based eukaryotic gene regulation will enable the facile construction of scalable synthetic gene circuits and open up new approaches for mapping natural gene networks and their effects on complex cellular phenotypes.

Characterizing and Alleviating Substrate Limitations for Improved in Vitro Ribosome Construction

Liu, Y., Northwestern University
Fritz, B. R., Northwestern University
Anderson, M. J., Northwestern University
Schoborg, J. A., Iowa State University
Jewett, M. C., Northwestern University

Title:

Characterizing and alleviating substrate limitations for improved *in vitro* ribosome construction

Authors:

Yi Liu, Brian R. Fritz, Mark J. Anderson, Jennifer A. Schoborg, Michael C. Jewett

Abstract:

We recently reported the development of an integrated synthesis, assembly, and translation (iSAT) method for *in vitro* construction of *Escherichia coli* ribosomes. iSAT allows simultaneous ribosomal RNA synthesis, ribosome assembly, and reporter protein expression as a measure of ribosome activity. Here we explore causes of iSAT reaction termination to improve efficiency and yields. We discovered that phosphoenolpyruvate (PEP), the secondary energy substrate, and nucleoside triphosphates (NTPs) were rapidly degraded during iSAT reactions. In turn, we observed a significant drop in the energy charge and termination of protein synthesis. Further, we identified that the accumulation of inorganic phosphate is inhibitory to iSAT. Fed-batch replenishment of PEP and magnesium glutamate (to offset the inhibitory effects of accumulating phosphate by repeated additions of PEP) prior to energy depletion prolonged the reaction duration 2-fold and increased yields 75%. By adopting a semi-continuous method, where passive diffusion enables substrate replenishment and byproduct removal, we prolonged iSAT reaction duration 5-fold and increased superfolder green
fluorescent protein yield 7-fold to 7.5 Â± 0.7 Âµmol L-1. This protein yield is the highest ever
reported for iSAT reactions. Our results underscore the critical role energy substrates play in iSAT and highlight the importance of understanding metabolic processes that influence substrate depletion for cell-free synthetic biology.

Bioprospecting for Genes That Confer Biofuel Tolerance Using a Genomic Library Approach

Tomko, T., The University of Vermont
Dunlop, M., The University of Vermont

Bioprospecting for Genes That Confer Biofuel Tolerance Using a Genomic Library Approach
Microorganisms are capable of producing advanced biofuels that can be used as â??drop-inâ?? alternatives to conventional liquid fuels. However, the cells often become overwhelmed by the toxic effects of the biofuel product. In order to make microbial biofuels a competitive fuel source, mechanisms of improving resistance to the toxic effects of biofuel production is vital. Our investigation aims to identify resistance mechanisms from microorganisms that have evolved to withstand extreme environmental pressures.
Using a plasmid-based transgenic library approach, genes believed to impart resistance can be inserted and studied in *Escherichia coli*. Adaptions such as increased efflux pump efficiency and less permeable cell membranes could help to improve biofuel tolerance in the host organism.

Pseudomonas aeruginosa was studied initially because the bacterium has evolved mechanisms to survive attacks from many damaging compounds. A plasmid library from P. aeruginosa was created and transformed into E. coli. The resulting cells were then stressed with biofuel to determine if any of the genetically altered E. coli displayed improved tolerance. Our study was successful in identifying a specific gene from P. aeruginosa that help to significantly increase tolerance to limonene when expressed in E. coli. The identified gene was an organic hydroperoxide resistance gene (ohr) that helped to alleviate the oxidative stress that limonene imposed on the cells.

Additionally, the transgenic library approach is being applied to other microorganisms such as *Alcanivorax borkumensis*. *A. borkumensis* has adapted to survive in harsh, hydrocarbon rich environments, very similar to the conditions encountered by a biofuel producing strain of bacteria. This library is being expressed in *E. coli* and tested to see if any biofuel tolerance genes can be recovered.

Directed Evolution of Far-Red Fluorescent Rhodopsins

McIsaac, R. S., Caltech
Engqvist, M., Caltech
Wannier, T.,
Herwig, L., Caltech
Arnold, F. H., California Institute of Technology

Microbial rhodopsins are a diverse group of photoactive trans-membrane proteins found in all three domains of life. A member of this protein family, Archaerhodopsin-3 (Arch) of halobacterium *Halorubrum sodomense*, was recently shown to function as a fluorescent indicator of membrane potential when expressed in mammalian neurons. Arch fluorescence, however, is very dim and is not optimal for applications in live-cell imaging. We used directed evolution to identify mutations that dramatically improve the absolute brightness of Arch, as confirmed biochemically and with live-cell imaging. At least some mutations discovered in one rhodopsin (GR) can also be transferred to another (Arch) to increase fluorescence. In several fluorescent Arch variants, the pKa of the protonated Schiff base linkage to retinal is near neutral pH, a useful feature for voltage sensing applications. These bright Arch variants also enable labeling of biological membranes in the far-red/infrared and exhibit the furthest red-shifted fluorescence emission thus far reported for a fluorescent protein.

Bottom-up Construction of Orthogonal Regulation for Gene Circuits Using Transcription Factor-Promoter Pairs with Predictable Properties

Blount, B. A., Imperial College London
Ellis, T., Imperial College London

The majority of well-described synthetic gene networks are constructed using the same transcription factors and so suffer from an inherent incompatibility. Although the number of transcription factors available to synthetic biology has expanded over recent years thanks to advances in modular zinc finger, CRISPR-Cas9 and TALE-based DNA binding proteins, generating and characterising unique synthetic transcription factor-promoter sets for every new network is time-consuming and impractical. Building on previous work where we introduced TAL Orthogonal Repressors (TALORs) for use in yeast, we show here that incorporating randomisation into TALOR assembly and coupling this with rapid promoter modification enables fast and efficient generation of valuable libraries of orthogonal transcription factor-promoter pairs. To do this we took a rational approach, first analysing a randomised 40-member constitutive promoter library using quantitative sequence-activity models to identify base positions within a core promoter that correlate with promoter output. This information was then applied to create design criteria for TALOR randomisation reactions; allowing partial randomisation of DNA-binding sites whilst maintaining output levels of all regulated promoters. In tests for orthogonality with 100 engineered strains, 10 regulated library promoters were evaluated against their 10 corresponding library TALORs, showing repression only when promoters are paired with their cognate library member. Remarkably, orthogonality was seen despite sequence homology in the 20 bp binding site between some members being as high as 85%. To demonstrate the functionality of this library, TALORs and their promoters were implemented as orthogonal wires in a novel yeast genetic circuit. Together, the methodology for generating libraries of orthogonal transcription factor-promoter pairs and the use of modelling for predictable properties potentially enables routine construction of synthetic circuits and networks from *de novo* regulatory components.

Exploiting Anaerobic Gut Fungi for Lignocellulose Breakdown and Enzyme Discovery

Solomon, K. V., University of California, Santa Barbara
Haitjema, C., University of California, Santa Barbara
Henske, J. K., University of California, Santa Barbara
Borges-Rivera, D., Broad Institute of Massachusetts Institute of Technology and Harvard
Thompson, D. A., Broad Institute of Massachusetts Institute of Technology and Harvard
O'Malley, M. A., University of California, Santa Barbara

Anaerobic gut fungi are attractive microbes to engineer for bio-based chemical production from lignocellulose, and are an untapped reservoir for the discovery of new cellulose-degrading enzymes. These fibrolytic microbes secrete an array of powerful cellulases and cellulolytic complexes (fungal cellulosomes) for synergistic hydrolysis of plant biomass, yet little sequence information exists for gut fungi. Though fungal hydrolytic activity has been shown to be substrate dependent, the underlying regulation mechanisms that coordinate the action of cellulases and cellulosomes from gut fungi remain unknown. We hypothesize that cellulose-degrading machinery is catabolite-repressed to conserve cellular energy, and our objective is to exploit this regulation mechanism for novel enzyme discovery. To address this hypothesis, we have combined next-generation sequencing and proteomic approaches to examine cellulose-degrading enzyme production in gut fungi while supported on several growth substrates. A new species of gut fungus from the *Piromyces* genus was isolated from the digestive tract of a horse, and its proliferation was monitored via fermentation gas production. Fungi exhibited high enzymatic reactivity against a range of cellulosic and lignocellulosic substrates (filter paper, Avicel, reed canary grass), which was repressed in the presence of simple sugars. Through strand-specific RNAseq and use of the TRINITY assembly platform, we were able to assemble hundreds of novel cellulase genes *de novo* from >27,000 transcripts without the need for genomic sequence information. The fungal transcriptome is particularly rich in GH6 and GH43-containing enzymes, and we find that 27 of 54 diverse glycosyl hydrolase families are transcriptionally repressed during growth on glucose relative to reed canary grass (lignocellulose). Within the majority of these transcripts, dockerin-tagged elements of fungal cellulosomes are abundant, and 15% of dockerin-containing transcripts are repressed in the presence of glucose. This suggests that catalytic components of fungal cellulosomes are highly regulated in response to simple sugars, which is supported by recently obtained proteomic data. By deciphering catabolic regulation patterns, we have identified several putative cellulose-active enzymes that are coordinately regulated with known cellulases through metatranscriptomic cluster analysis. These transcripts bear no homology to known glycosyl hydrolases, and are potentially completely divergent novel enzymes. We will further discuss the transcriptional regulation patterns observed for other important enzyme families under catabolic regulatory conditions, and connect these regulation patterns to protein secretion and lignocellulose degradation across gut fungal genera.

A Genetically-Structured Deterministic Chemical Kinetic Simulation of the Life Cycle of the Biotechnologically Important Filamentous Bacteriophage M13

Fisk, J.,
Smeal, S.,
Prasad, A., Colorado State University

Bacteriophages have been central to the development of molecular biology, serving as model systems for understanding molecular phenomena and as components of important biotechnologies. Phage systems remain at the forefront of research in nanomaterials and nanomedicine and are widely employed component-generating technologies for drug and materials discovery, primarily through phage display. We describe the construction of a genetically-structured, experimentally-based computational simulation of the life cycle of the biotechnologically important filamentous bacteriophage M13. Our simulation integrates 50 years of experimental observations to evaluate and expand the system level understanding of M13. Our deterministic chemical kinetic simulation explicitly includes the molecular details of phage DNA replication, mRNA transcription, protein translation and phage particle assembly, as well as the competing protein-protein, protein-double-stranded DNA, protein-single-stranded DNA, and protein-mRNA interactions that control the timing and extent of phage production. Many aspects of M13 biology are faithfully reproduced by the simulation, including the production levels and timing of the shift between replicative form and preassembly complex DNA, quantities of phage mRNA and proteins, and the time course of assembly and release of progeny phage. The simulation was employed to evaluate the importance of various inherent and dynamic elements in the regulation of the phage life cycle. We describe the relative importance of phage genome structure, promoter and ribosome binding site strengths, and dynamic protein and mRNA levels and binding interactions in controlling the initial phage infection. The simulation that was constructed to describe the course of infection in a single cell over one cell cycle was expanded to provide a picture of the molecular interactions involved in establishing the dynamic steady states that define the persistently infected state of a cell across several cell generations. Our simulation provides a quantitative description of phage biology and highlights gaps in the present understanding of the M13 life cycle. We expect that the simulation will help prioritize future biochemical experiments and be a useful tool in designing new synthetic phage-like systems.

Engineering Transcriptional Regulator Effector Specificity through Rational Design and Rapid Prototyping

de los Santos, E. L. C., California Institute of Technology
Meyerowitz, J. T., California Institute of Technology
Mayo, S. L., California Institute of Technology
Murray, R. M., California Institute of Technology

The pursuit of circuits and metabolic pathways of increasing complexity and robustness in synthetic biology will require engineering new regulatory tools. Feedback control based on relevant molecules, including toxic intermediates and environmental signals, would enable genetic circuits to react appropriately to changing conditions. In this work, computational protein design was used to create functional variants of qacR, a tetR family repressor, responsive to a new targeted effector. The modified repressors target vanillin, a growth-inhibiting small molecule found in lignocellulosic hydrolysates and other industrial processes. A computationally designed twenty-seven member library was screened using an *in vitro* transcription-translation (TXTL) system. Leads from the *in vitro* screen were characterized *in vivo*. Preliminary results demonstrate dose-dependent regulation of a downstream fluorescent reporter by vanillin. These repressor designs provide a starting point for the evolution of improved variants. We believe this process can serve as a framework for designing new sensors for other target compounds.

Fatty Acid Responsive Hybrid Promoter Design in Oleaginous Yeast Yarrowia Lipolytica

Shabbir-Hussain, M., Clemson University
Blenner, M. A., Harvard Medical School & Children's Hospital Boston

Fatty Acid Responsive Hybrid Promoter Design in Oleaginous Yeast Yarrowia lipolytica

Murtaza Shabbir Hussain[1], Samuel Williams[1], **Mark Blenner**[1]

[1]Department of Chemical & Biomolecular Engineering, Clemson University, Clemson, SC 29634.

Yarrowia lipolytica is a non-conventional oleaginous yeast that grows on numerous substrates, including glycerol, fats, and other hydrophobic carbon. As an efficient producer of citric acid, lipase, and fatty acids, Y. lipolytica has gained attention as a bioproduction organism. Well- controlled metabolic engineering of this yeast requires a set of inducible promoters that specifically regulate gene expression levels in response to specific molecules within the cell as well as in the extracellular environment. As an oleaginous yeast, the engineering of fatty acid synthesis and metabolism is of great interest, and considerably little information is available on the molecular details of fatty acid regulated gene expression in Y. lipolytica. Through a series of promoter truncations of two fatty acid responsive genes, acyl-CoA oxidase 2 (POX2), and acetyl-CoA acetyl transferase (PAT1), we have identified key elements in fatty acid regulated promoters including upstream activating sequences (UAS) and core promoter elements that confer fatty acid responsiveness. We measured gene expression levels using real-time PCR and GFP expression, and studied the effect of different fatty acids on gene expression. We provide a quantitative relationship between fatty acid concentration and gene expression. Furthermore, we describe the uncoupling of fatty acid responsive promoters from catabolite repression. Combining this information, we are creating new synthetic hybrid promoters of well-defined strength that have well-defined fatty acid response profiles. With new well-defined inducible promoters, metabolic engineering of fatty acid regulated pathways will increase yield, productivity, and the scope of molecules produced by Y. lipolytica.

Sensitivity of Synthetic Biological Circuits to Environmental Conditions

Venayak, N.,
Mahadevan, R., University of Toronto

The field of synthetic biology enables the design of complex genetic circuits which can be adapted to solve technical challenges in a wide variety of disciplines. To facilitate rapid engineering, it is desirable for these systems to be modular and transferable. Currently, however, their characterization in varying environmental conditions is limited, with most synthetic circuits having been developed and characterized under single, ideal conditions. In metabolic engineering and industrial biotechnology, the diversity of the products, organisms, and environments necessitates systems that are well characterized, functional, and robust towards changes in these variables. In this work, we determine the functionality of different implementations of the genetic toggle switch, one of the seminal feats of synthetic biology. By exposing different toggle switches to varying environmental conditions, we have shown that these systems are heavily dependent on their environment, limited not only to media composition but also oxygen availability. In addition to these conditions affecting the efficiency of these systems, they can also fully inactivate them, causing them to behave unpredictably. An understanding of these synthetic biological circuits and their activity in different environments is crucial to ensure their modularity, while increasing their complexity and allowing their adoption and evolution within new fields.

A Highly Tunable System for the Simultaneous Expression of Multiple Enzymes in Saccharomyces Cerevisiae

Ito, Y., TOYOTA Central R&D Labs., Inc.
Yamanishi, M., TOYOTA Central R&D Labs., Inc.
Ikeuchi, A., Toyota Central R&D Labs., Inc.
Matsuyama, T., TOYOTA Central R&D Labs., Inc.

Control of the expression levels of multiple enzymes in transgenic yeasts is essential for the effective production of complex molecules through fermentation. Here, we propose a tunable strategy for the control of expression levels based on the design of terminator regions and other gene-expression control elements in *Saccharomyces cerevisiae*. Our genome-integrated system, which is capable of producing high expression levels over a wide dynamic range, will broadly enable metabolic engineering and synthetic biology. We demonstrated that the activities of multiple cellulases and the production of ethanol were doubled in a transgenic yeast constructed with our system compared with those achieved with a standard expression system.

Rapid Evaluation of Itaconic Acid Production Strategies in Saccharomyces Cerevisiae

Wu, L., DSM Biotechnology Center, Alexander Fleminglaan 1, 2613 AX Delft, The Netherlands
Zhao, Z., DSM Biotechnology Center
Meijerinck, B., DSM Biotechnology Center
Gielesen, B., DSM Biotechnology Center
Ozalp, B., DSM Biotechnology Center
van der Hoeven, R., DSM Biotechnology Center
Bovenberg, R., DSM Biotechnology Center
Roubos, H., DSM Biotechnology Center

The recent explosion of strain optimization algorithms based on genome-scale metabolic models (GSMM) allows for rapid generation of a large number of metabolic engineering strategies for maximizing metabolite production. However, implementation and validation of these strategies remains a bottleneck in the design-build-test-learn cycle. Here we demonstrate a method for targeted high-throughput strain transformation using exchangeable building blocks. Using this method and a high-throughput analysis platform, we evaluated multiple itaconic acid producing strategies simultaneously in *Saccharomyces cerevisiae* wild type strains. The initial design included differences in compartmental strategy, a set of gene variants, differences in promoter strength and two strain backgrounds. The results revealed further potential for improving the productivity of itaconic acid. Overall, we demonstrate that rapid prototyping by targeted engineering has become a valuable tool to quickly explore metabolic engineering scenario's including host strains for the production of a heterologous product.

Engineering and Understanding Enzyme and Pathway Localization to ER and Lipid Droplet Membranes

Lin, J. L., UC Riverside
Wheeldon, I., UC Riverside

Many essential and industrially important metabolic pathways are assembled on cellular membranes: proteins are glycosylated through a series of membrane-bound reactions; in plants isoprenoids are synthesized in well-organized multienzyme structures bound to intracellular membranes; and, the key step in acetate ester production in yeast is localized to the membrane surface of lipid droplets. Such localization is advantageous. High local concentrations of enzymes and reaction intermediates can increase pathway reactions rates and prevent undesired side reactions from consuming pathway intermediates, thus leading to increased pathway yields. Intracellular scaffolding technologies (e.g., protein and nucleic acid scaffolds) have re-created these advantages in engineered pathways by co-localizing enzymes in the cytosol. While these technologies have produced substantial gains in pathway yields they cannot readily accommodate membrane-bound enzymes. This gap in synthetic biology tools significantly limits our abilities to access the benefits of enzyme co-localization with membrane-bound pathways.

The goal of this work is to identify protein parts that can be used to direct the assembly to biosynthetic pathways comprised of natively membrane-bound and cytosolic enzyme on intracellular membranes. Specifically, we aim to develop synthetic biology tools that can sort endoplasmic reticulum (ER) enzymes to lipid droplets (LDs) and that can direct cytosolic enzyes to LDs. In this way, we aim to create co-localized reaction centers in the intracellular space. The first step in engineering such a system is to understand and control localization to LDs. For example, alcohol acetyltransferase-1 in *S. cerevisiae*, which catalyzes the condensation of simple alcohols and acetyl-CoA to the corresponding ester, sorts from the ER to lipid droplets where it remains during stationary phase. Through a series of structure-function studies we have identified dual N- and C-terminal amphipathic helices that are responsible for this intracellular localization. Our studies of LD localization extend to a series of protein tags known to function in mammalian and plant cells. We demonstrate that the terminal hydrophobic domain from a putative methyltransferase AAM-B can localize cytosolic proteins to LDs via the ER. Under fermentation conditions, we demonstrate that the LD protein olesin can be used as a terminal fusion to localize proteins to LDs in yeast. Building from these localization studies we aim to organize LD localized multistep reaction pathways for the synthesis of acetate and ethyl esters.

An Orthogonal DNA Replication System in Yeast

Ravikumar, A., UC Irvine
Arrieta, A., Biomedical Engineering
Liu, C. C., University of California at Irvine

An extranuclear replication system, consisting of an orthogonal DNA plasmid–DNA polymerase pair, was developed in Saccharomyces cerevisiae. Engineered error-prone DNA polymerases showed complete mutational targeting in vivo: per-base mutation rates on the plasmid were increased substantially and remained stable with no increase in genomic rates. Orthogonal replication serves as a platform for in vivo continuous evolution and as a system whose replicative properties can be manipulated independently of the host's. Specifically, its application towards (1) the study of the reversibility of enzyme evolution, and (2) the evolution of global transcription factors for improved yeast tolerance to environmental stresses (e.g. high levels glucose and ethanol, low pH, and high heat), are discussed.

Engineering Protein-Nucleic Acid Binding for Cellular Circuits

Mansell, T. J., University of Colorado
Domènech Corts, A., University of Colorado
Gill, R. T., University of Colorado

Cellular circuits in synthetic biology often depend on inducible regulators like the lac and tet repressors. However, increased circuit complexity is limited by the availability of orthogonal regulators. We present a method for generation of directed libraries of regulators for binding of DNA and RNA sequences in *E. coli*. Our unique logic-based selection strategy allows for selection of repressors that do bind a target (i.e., novel) sequence but do not bind a pre-determined sequence of interest (e.g., wild type). We leverage the technologies of multiplex oligo synthesis, recombineering, and next-generation sequencing to create libraries and glean sequence-activity relationship data from selections to iterate evolution in a rational, directed way.

Construction of a Metabolite Valve in Saccharomyces Cerevisiae Increases Pathway Yields

Tan, S. Z., MIT
Prather, K. L. J., Massachusetts Institute of Technology

Background: Microbial fermentation has the potential to produce various chemicals from simple and inexpensive starting materials that are otherwise derived from non-renewable or limited resources. Typical methods to improve product yields include overexpressing pathway enzymes and deleting competing pathways. However, these methods are constrained by the inter-connectedness of metabolic pathways and the finite nature of cellular resources. Overexpression of pathway enzymes comes in the expense of endogenous ones due to consumption of common precursors. On the other hand, deletion of competing pathways may necessitate media supplementation and impair cellular function. **Methods:** To overcome this, we constructed of a metabolite valve that has the potential to improve pathway yields without affecting cellular function. A metabolite valve enables (1) down-regulation of competing pathway enzyme(s) and (2) dynamic control of these enzyme levels. As a proof-of-concept, we looked at directing metabolic flux away from glycolysis into a model pathway, gluconate. In a *Saccharomyces cerevisiae* strain where Glucokinase 1 and Hexokinase 2 have been deleted, a metabolite valve was constructed at the only remaining hexokinase, Hexokinase 1. The promoter of Hexokinase 1 was replaced with the inducible tetracycline transactivator protein (tTA) system, and glucose dehydrogenase from *B. subtilis* was expressed for the production of gluconate. **Results:** Regulation of Hexokinase 1 expression at the transcription level resulted in a maximum 10-fold decrease in enzyme activity upon induction with doxycycline. This reduction in enzyme activity resulted in an increase in available glucose for the gluconate pathway. Consequently, a 300-fold increase in yield of gluconate from glucose was observed, from 0.1 mol/mol % to 18 mol/mol % upon induction with doxycycline. Moreover, a 30% increase in biomass yield was observed; repressing Hexokinase 1 with the valve resulted in increased carbon efficiency and less ethanol production. **Conclusions:** This proof-of-concept presents an innovative strategy to increase product yields by controlling competing native metabolic pathways. With the success of this proof-of-concept, we would like to implement a metabolite valve to improve the yield of glucaric acid, a "top 10 value added chemical" listed by the USDA. Here, multiple valves can be constructed at competing enzymes such as PFK1, PFK2, and PIS1. In addition to control at the transcription level with the existing tTA system, post-translational methods such as controlled protein degradation and the CRISPR Interference system will be explored.

Synthetic Approaches to Investigate Intracellular Transport and Cytoskeletal Regulatory Systems

Zimak, J., Rice University
Tsao, D., Rice University
Kumar, E., Rice University
McLaughlin, T., Rice University
Diehl, M., Rice University

Eukaryotic cell physiology is deeply dependent on the spatial control of intracellular chemistry. The regulation of cytoskeletal filament architecture and the activated transport of material within these networks therefore have a profound influence over a wide range of biophysical processes and cell signaling responses. Moreover, the development of methods to modulate cytoskeletal structure and intracellular transport are uniquely important to gaining synthetic control over many eukaryotic cell functions. Herein, we describe synthetic methods to engineer activated intracellular transport behaviors in living cells. This is accomplished via the genetic-level programming of various molecular motors, their coupling to vesicular cargos and cargo sizes. We also demonstrate how this procedure can be adapted to examine functional relationships among multiple interacting regulatory proteins that influence cytoskeletal architecture, cell morphology and cell-cell adhesion. In each case, these methods offer opportunities to dissect fundamental dynamic responses of coupled macromolecular systems and characterize their collective response to specific regulatory factors and perturbations.

The Development of in Vitro-Based Biosensors for the Detection of Pathogenic Biofilms

Wen, K. Y., Imperial College London
Chappell, J., Cornell University
Jensen, K., Imperial College London
Freemont, P. S., Imperial College London
Filloux, A., Imperial College London
Davies, J., 5Royal Brompton & Harefield NHS Foundation Trust

The development of *in vitro*-based biosensors for the detection of pathogenic biofilms

Ke Yan Wen1,2, James Chappell3, Kirsten Jensen1,2, Jane Davies4,5, Alain Filloux6, Paul
Freemont1,2

1Centre for Synthetic Biology and Innovation, Imperial College London, UK
2Department of Life Sciences, Imperial College London, UK
3School of Chemical and Biomolecular Engineering, Cornell University, Ithaca, NY, USA
4Department of Gene Therapy, Imperial College London, UK
5Royal Brompton & Harefield NHS Foundation Trust, London, UK
6Medical Research Council Centre for Molecular Bacteriology and Infection, Department of Life Sciences, Imperial College London, UK

Microbial biofilms are cooperative cell communities that coordinate behaviour and gene expression at the population level. They are extremely difficult to disperse with antimicrobial agents due to their ability to produce a protective extracellular matrix. Subsequently they
pose a major challenge in healthcare and industrial settings. For example, cystic fibrosis patients often suffer chronic lung infections that are due to biofilm colonisations of *Pseudomonas aeruginosa.* Rapid detection of this bacterial phenotype could assist greatly in preventing and treating the establishment of chronic infections. We have developed an *in vitro* biosensor to detect the presence of *P.aeruginosa* quorum-sensing molecules. It is based on the expression of a DNA template within a cell-free system, to produce both the detection module and a measurable output in a single reaction. Our use of cell-free coupled
transcription and translation systems as a biosensor platform provides an alternative to whole- cell chassis, which offers advantages in terms of simplicity, detection time and biosafety.

After establishing a proof of principle for the detection of the presence of *P. aeruginosa* biofilms using this biosensor, we are currently focused on testing clinical samples from cystic fibrosis patients. This approach will allow us to determine the potential for this method as a research and diagnostic tool for pathogenic biofilms.

Quantitative Prediction of Therapeutic Fusion Protein Dynamics

Robinson-Mosher, A., Harvard Medical School
Chen, J. H., Harvard Medical School
Way, J., Harvard University
Silver, P. A., Harvard Medical School

Design tools for synthetic biological devices are not yet to the point where they can provide reliable predictions even for simple systems. We address this problem for the case of fusion proteins engineered for therapeutic targeting. In particular, we address fusions consisting of two well-characterized domains separated by a linker, where one domain binds to a target epitope and the second exhibits desired biological activity. A two-component modeling approach lets us make quantitative predictions that can be checked against our experimental results to validate the success of the model. The model is composed of a constrained Brownian dynamics (CBD) simulation component that predicts the effect of linker configuration on fusion protein activity and an ordinary differential equation component that uses details of the experimental setup, kinetic information about the protein domains and results from the CBD component to generate predicted dose-response curves. We demonstrate application of the combined model to our therapeutic constructs via *in vitro* cell-based assays. We observed a linker length-dependent effect on targeting efficiency, as well as a moderate effect due to modulating the binding strength of the biologically active domain. We also generated a two-dimensional landscape mapping binding strength to predicted targeting efficiency and observed the existence of a plateau of maximal targeting which falls off steeply at high binding strengths. The ability to predict optimal values for targeting of both linker configuration and binding strength will allow us to rationally generate candidate therapeutic designs.

Programmable Bacteria As Diagnostics of Gut Health

Kotula, J., Harvard University

The mammalian gastrointestinal (GI) tract contains a dynamic community of symbiotic microbes that interact with the host to impact health, disease and metabolism. The gut microbiota are uniquely positioned to probe the GI environment, because they interact closely with the intestinal lumen as well as the host immune system. We constructed engineered bacteria to sense, remember, and report on the environmental conditions of the mammalian gut. Based on previous genetic memory systems, we constructed a modular, two-part system with a 'trigger element' that expresses the lambda cro gene in response to an environmental signal, and a 'memory element' derived from the cI/Cro transcriptional switch found in phage lambda. The memory element has an extremely stable cI state and a Cro state that is stable for many cell divisions. When engineered *E. coli* were administered to mice, and treated with anhydrotetracycline to trigger the memory element, all of the recovered *E. coli* switched from the cI state to the Cro state. However, those recovered from the untreated mice remained completely in the cI state. The trigger and memory elements were transferred from *E. coli* K12 to an uncharacterized *E. coli* strain isolated from the mouse. This newly engineered strain triggered in response to and maintained memory of stimulus exposure in vitro and within the mouse GI tract. Additionally, the engineered murine *E. coli* maintained a stable population within the mouse gut. Our engineered elements are very modular, therefore the probiotics can be easily modified to sense various diseases, such as Crohn's colorectal cancer, or the pathogens that cause *C. diff* infections as well as traveller's diarrhea. Therapeutic elements can also be added that would alleviate disease symptoms, or specifically kill pathogenic bacterial strains. This work demonstrates that *E. coli* can be engineered into living diagnostics capable of noninvasively probing the mammalian gut, and further lays a foundation for the use of synthetic genetic circuits as monitoring systems in complex, ill-defined environments.

Controlled Measurements of Multiple Cellular Components in E. coli As a Resource for Integrative Computational Modeling of Cellular Subsystems

Houser, J., Univ Texas
Barnhart, C.,
Boutz, D.,
Carroll, S.,
Michener, J. K., Harvard University
Needham, B.,
Papoulas, O.,
Sridhara, V.,
Sydykova, D.,
Marx, C. J., University of Idaho
Trent, M. S.,
Barrick, J. E., University of Texas at Austin
Wilke, C.,
Marcotte, E.,

Genome wide models are a useful tool in synthetic engineering of biology. Most genome-wide models however focus on individual subsystems, or networks, of the cell, such as in flux balance models of metabolism. Recent effort has been focused on how to best integrate models of separate subsystems (e.g. transcription, metabolism, etc.). Thus, there is a need for consistent and comparable measurements of these different cellular components. Though many large data sets exist for specific types of cellular components (e.g. RNA expression, protein abundances, metabolites, etc.) there is a lack of data that span multiple different sub-systems under the same controlled conditions. Here we present genome-wide measurements on RNA and proteins as well as lipids and metabolic fluxes from *E. coli* grown under identically controlled conditions. These data serve as a resource for building integrative models of cellular processes. Specifically, we present data from a long-term glucose starvation time course. These data comprise the first completed results from a series of planned experiments considering different environmental conditions. We will discuss the general trends seen in this time course as well as compare and contrast our data sets on different sub-processes. Finally, we will present a mechanistic, large-scale, and predictive model of gene regulation connecting protein and RNA expression profiles. This model could be used to quantitatively predict changes in protein expression given differences in genetic background and/or environmental conditions with a minimum of additional information.

Improving Polyketide Production By Screening Heterologously Expressed Synthetic Propionyl-CoA Carboxylases

Vandova, G., Stanford University
O'Brien, R., Stanford University
Harvey, C., Stanford University
Lowry, B., Stanford University
Davis, R., Stanford Genome Technology Center
Khosla, C., Stanford University
Hillenmeyer, M., Stanford University

Polyketides are valuable natural products that have antibacterial, anticancer, and immunosuppressant activities. Many bacterial polyketides are produced by modular polyketide synthases (PKS) in an assembly line fashion from simple two-carbon monomer inputs. Two PKS pathways, encoding erythromycin and epothilone, have been heterologously expressed in *Escherichia coli* (*E. coli*), but at low yields (0.001-20mg/L). We aimed to improve *E. coli* as a host for polyketide production by providing more monomer inputs for PKS assembly lines. For many polyketides, the extender monomer is methylmalonyl-CoA and the enzyme that provides it is propionyl-CoA carboxylase (PCC). We used as our test system a strain of *E. coli* harboring the 6-deoxyerythronolide B synthase PKS, which produces the erythromycin polyketide. We designed and synthesized DNA encoding 20 PCC gene homologs from different species and built 20 versions of the *E. coli* polyketide-producing strain, each harboring a different PCC homolog. We observed variability in polyketide production amongst the 20 strains, and identified optimal PCC homologs. These results may be useful for improving *E. coli* as a host for heterologous expression of orphan PKSs, as well as for engineering polyketides and producing novel unnatural products.

Characterization of the Performance of Non-Fluorescent Genetic Constructs Via Indirect Competitive Construct Characterization (iCCC)
Hecht, A., National Institute of Standards and Technology
Munson, M. S., National Institute of Standards and Technology
Endy, D., Stanford University
Salit, M., National Institute of Standards and Technology

Characterization of the Performance of Non-Fluorescent Genetic Constructs *via* Indirect

Competitive Construct Characterization (iCCC)

Ariel Hecht, Matthew S. Munson, Drew Endy, and Marc Salit

Efficient engineering of complex metabolic pathways into organisms depends on the ability to predict the function of a composed system from knowledge of its constitutive parts. Our work is focused on the development of methods for parts characterization that yield reproducible results that can be meaningfully compared across different experiments, different labs, and different times. This capability could yield a database of engineering reference data that can be used with confidence to support design, reusability, and predictive modeling.

Current methods for characterizing circuit performance require including a fluorescent protein gene in the construct of interest, which can significantly alter its behavior.1,2 To overcome this limitation, we are developing the indirect Competitive Construct Characterization (iCCC) assay as a method for quantifying the performance of any non-fluorescent genetic construct.

We propose an assay system where the construct of interest is expressed in competition with a reference construct producing a fluorescent protein in a resource-limited environment. The use of a cell free expression system allows for engineering control over the resources available for transcription and translation, allowing the environment to be tailored to ensure competition between the constructs. The ratio of fluorescent protein produced under competitive and non- competitive conditions, should be a robust measure of the activity of the part of interest. The key assumptions underlying this method are:

(1) Consumption of resources by a construct is a meaningful measure of its activity

(2) Resource limitations affect the activity of both constructs in a known manner

(3) In a resource limited environment, consumption of resources by the construct of interest results in reduction of protein production by the reference construct
We have adapted and applied mathematical models3â??5 of transcription and translation of two constructs coupled through their utilization of shared resources (NTPs, amino acids, polymerases, and ribosomes). We will show the use of this model to explore parameter space allowing determination of regimes under which competition for resources can be most meaningfully observed as a decrease in the production of the reference construct. We use this to provide definition of the potential operating limits of the assay, as well as guide design of the characteristics of useful reference constructs.

References

1. Mutalik, V. K. *et al.* Precise and reliable gene expression via standard transcription and translation initiation

elements. *Nat. Methods* **10,** 354â??60 (2013).

2. Mutalik, V. K. *et al.* Quantitative estimation of activity and quality for collections of functional genetic elements. *Nat. Methods* **10,** 347â??53 (2013).

3. Siegal-gaskins, D., Noireaux, V. & Murray, R. M. Biomolecular resource utilization in elementary cell-free gene circuits. *Am. Control Conf. (ACC), 2013* 1531â??1536 (2013).

4. StÃ¶gbauer, T., Windhager, L., Zimmer, R. & RÃ¤dler, J. O. Experiment and mathematical modeling of gene expression dynamics in a cell-free system. *Integr. Biol.* **4,** 494â??501 (2012).

5. Niederholtmeyer, H., Stepanova, V. & Maerkl, S. J. Implementation of cell-free biological networks at steady state. *Proc. Natl. Acad. Sci. U. S. A.* **110,** 15985â??90 (2013).

Harnessing Peroxisomes for Metabolic Engineering

Russ, Z. N., UC Berkeley
DeLoache, W. C., UC Berkeley
Dueber, J. E., U.C. Berkeley

Membrane-bound organelles have started to interest synthetic biologists as a convenient framework for scaffolding and compartmentalizing metabolic pathways. One such organelle, the peroxisome, offers ease of use and large capacity for protein import, but thorough investigation and characterization is required before the peroxisome can be repurposed as a "synthetic organelle." We have examined both peroxisomal protein import and small metabolite permeability, developing novel *in vivo* assays to measure both aspects. We have also applied these assays to develop improved peroxisomal targeting signals in a high-throughput manner. Finally, we show how these enhanced tags can be used to sequester pathways and engineer an ideal synthetic organelle strain.

High-Throughput Crispri Studies of Gene Function and Regulation in E. coli

Toro, E., University of California Berkeley
Rishi, H. S., Physical Biosciences Division
Chen, D., University of California Berkeley
Cambray, G., California Institute for Quantitative Biosciences, University of California, Berkeley
Wang, X., Tsinghua University
Liu, H., Tsinghua University
Qi, L. S., California Institute for Quantitative Biomedical Research
Arkin, A. P., University of California at Berkeley

The CRISPR-associated protein Cas9 has been adapted as a versatile tool for transcriptional regulation, genome editing, and imaging in a number of organisms. Here, we apply the catalytically inactive dCas9 to conduct high-throughput transcriptional and regulatory studies in *E. coli.* Using an Agilent OLS library of 32992 unique sgRNAs, we targeted 4500 genes, 5400 promoters, 640 transcription factor binding sites, and 106 sRNAs in the E. coli genome. These genomic targets cover a wide range of functionalities such as metabolism, stress response, transport, and cell division. By leveraging CRISPRi with next-generation sequencing, we were able to interrogate the fitness effect of transcriptional knockdown for each of the aforementioned genomic features in a single-pot experiment both aerobically and anaerobically. Our fitness results agreed well with current knockout databases, and our ability to induce transcriptional knockdown at any point during an experiment has allowed us to explore target essentiality under different conditions (aerobic, anaerobic, etc.) with great ease. We demonstrated this by showing that although *nrdA* and *nrdB* are essential under aerobic conditions â?? and are annotated as such in databases â?? they are dispensable anaerobically. Furthermore, our preliminary screens have identified the most effective sgRNAs for targeting E. coli features and have provided useful data on what makes different sites and sgRNA designs most effective. Overall, HT-CRISPRi enables single-pot, precise measurements of fitness for a large set of genomic features and will prove useful in genomic studies of model and non-model organisms.

Model-Driven Engineering of Gene Expression from RNA Replicons

Beal, J., Raytheon BBN Technologies
Wagner, T.,
Kitada, T.,
Azizgolshani, O.,
Moberg Parker, J.,
Densmore, D.,
Weiss, R., M.I.T.

RNA replicons are an emerging platform for engineering synthetic biological systems. Replicons self-amplify, can provide persistent high-level expression of proteins even from a small initial dose, and, unlike DNA vectors, pose minimal risk of chromosomal integration. However, no quantitative model sufficient for engineering levels of protein expression from such replicon systems currently exists. Here, we aim to enable the engineering of multigene expression from more than one species of replicon by creating a computational model based on our experimental observations of the expression dynamics in single- and multireplicon systems. To this end, we studied fluorescent protein expression in baby hamster kidney (BHK-21) cells using a replicon derived from Sindbis virus (SINV). We characterized expression dynamics for this platform based on the dose-response of a single species of replicon over 50 h and on a titration of two cotransfected replicons expressing different fluorescent proteins. From this data, we derive a quantitative model of multireplicon expression and validate it by designing a variety of three-replicon systems, with profiles that match desired expression levels. We achieved a mean error of 1.7-fold on a 1000-fold range, thus demonstrating how our model can be applied to precisely control expression levels of each Sindbis replicon species in a system.

Extrinsic Noise and Low-Level Pulsing Orchestrate Bet-Hedging Responses
Garcia-Bernardo, J.,
Dunlop, M., The University of Vermont

Extrinsic Noise and Low-Level Pulsing Orchestrate

Bet-Hedging Responses

Javier Garcia-Bernardo1*, Mary J. Dunlop2

1 Department of Computer Science, University of Vermont, 301 Votey Hall, Burlington, VT 05405

2 School of Engineering, University of Vermont, 301 Votey Hall, Burlington, VT 05405

* Corresponding author: jgarciab@uvm.edu

ABSTRACT

Cells live in changing, uncertain environments, exposed to sudden increases in the concentration of stressors. In order to survive, cells cannot depend solely on sensory responses, which have a delay associated with them, instead they need to anticipate future changes. However, the continuous synthesis and maintenance of general stress response mechanisms has a high energy cost for the cell. To resolve this, cells can rely on phenotypic diversity, where only a few cells carry the burden of expressing resistance mechanisms, effectively bet-hedging against the sudden appearance of a stressor. Phenotypic diversity can be produced by fluctuations in gene expression and is observed in many stress response mechanisms. In this study, we show another role of noise in gene expression. By simulating a transcription factor that activates several downstream genes (single-input module) we discovered that even when the concentration of the regulator is kept at low levels, infrequent activation of the downstream genes is observed. Expression of the downstream genes can be coordinated in the presence of extrinsic noise or low-level upstream dynamics. We compared an activator with fixed expression to an activator with pulsing dynamics. With one downstream gene, there is little or no difference observed between the two activator profiles. However, when several downstream genes are studied together, the pulsing activator is able to coordinate the response, while maintaining the same cost for the cell. Furthermore, this cost can be tuned by modifying the frequency, duration and amplitude of the pulses of the activator, providing the cell with a flexible mechanism for tuning the dynamics. A similar effect is seen when all the downstream genes are subject to extrinsic noise in their expression. This coordination may not be exclusive to stress response mechanisms and has the potential to provide an advantage in all single-input modules where coordination and low levels of the regulator is desired.

The Layered Analysis and Control of Biochemical Reaction Networks
Prescott, T. P.,
Papachristodoulou, A., Oxford University

Layered analysis and control of biochemical

reaction networks

Thomas P. Prescott and Antonis Papachristodoulou Department of Engineering Science, University of Oxford, United Kingdom E-mail: thomas.prescott@dtc.ox.ac.uk

Abstract

The conceptual hierarchy of biological parts and modules 1 is an important design principle for engineering synthetic biological systems, or enhancing existing systems to have additional functional sophistication. Many researchers have suggested that the modular design princi- ple is mirrored in evolved biological systems, and a large amount of research has uncovered modular structures in large-scale biological systems. 2â??5 By decomposing such a system, its analysis and redesign can be decomposed into that of its subsystems and their interconnec- tion. Modules are a specific type of subsystem made up of groups of biochemical species which together perform a specific function. However, the mechanisms by which modules in- teract often result in retroactivity, 6 where the behaviour of isolated modules changes in the context of the overall system. This impedance effect can cause synthetic modules to behave differently to expectations. We propose an alternative approach to the characterisation of network structure, termed layering. 7,8 Subsystems (called layers) are defined as groups of reactions, where any number of layers can contribute to the evolution of each species. In existing networks, strategies for choosing the reaction partition include taking separation in

?To whom correspondence should be addressed

1
timescale or spatial scale into account. An important consequence of the layering framework for Synthetic Biology is that it distinguishes retroactivity, which arises from overlaying mul- tiple layersâ?? contributions to the same species, from inter-layer dependencies. By making this distinction, the context of each synthetic layer is treated as an input, and its inputâ??output behaviour therefore remains the same in any context. Thus layering presents a flexible new design principle for engineering synthetic

biological systems.

Keywords Biochemical reaction networks; Structure; Modularity; Retroactivity; Layers

References

(1) Slusarczyk, A. L.; Lin, A.; Weiss, R. *Nature Reviews Genetics* **2012**, *13*, 406â??420.
(2) Alexander, R. P.; Kim, P. M.; Emonet, T.; Gerstein, M. B. *Science Signalling* **2009**, *2*, pe44.
(3) Kholodenko, B. N.; Kiyatkin, A.; Bruggeman, F. J.; Sontag, E.; Westerhoff, H. V. *PNAS*
2002, *99*, 12841â??12846.
(4) Saez-Rodriguez, J.; Kremling, A.; Conzelmann, H.; Bettenbrock, K.; Gilles, E. D. *IEEE Control Systems Magazine* **2004**, *24*, 35â??52.
(5) Milo, R.; Shen-Orr, S.; Itzkovitz, S.; Kashtan, N.; Chklovskii, D.; Alon, U. *Science* **2002**,

298, 824â??827.

(6) Del Vecchio, D.; Ninfa, A. J.; Sontag, E. D. *Molecular Systems Biology* **2008**, *4*, 161.
(7) Prescott, T. P.; Papachristodoulou, A. Layering in networks: The case of biochemical systems. 2013.
(8) Prescott, T. P.; Papachristodoulou, A. Signal propagation across layered biochemical networks. 2014; In Press.

Ultra-High-Throughput Screening of Enzyme Libraries with Droplet-Based Microfluidics

Romero, P., University of California, San Francisco
Tran, T., University of California, San Francisco
Abate, A. R., California Institute for Quantitative Biosciences, University of California, San Francisco

Ultra-high-throughput screening of enzyme libraries with droplet-based microfluidics

Philip A. Romero, Tuan M. Tran, and Adam R. Abate

Department of Bioengineering and Therapeutic Sciences, University of California, San

Francisco, 1700 4th St., San Francisco, CA 94158

Directed evolution relies on the ability to screen large, diverse libraries of sequence variants. We have developed a microfluidic platform for ultra-high- throughput screening of enzyme libraries. Individual enzyme variants are assayed in a monodisperse water-in-oil microemulsion. Aqueous droplets compartmentalize the enzymatic reaction and maintain the genotype-phenotype linkage. A quantitative fluorescence detection system allows droplets to be analyzed and sorted at rates beyond 1 kHz. This screening platform provides a throughput that is orders of magnitude beyond traditional plate-based screens.

Construction and Characterization of Synthetic Dual-Input Promoters

Roney, I., University of Ottawa
Kærn, M., Ottawa Institute of Systems Biology

Many genes are regulated at the level of transcriptional initiation by the interaction of multiple transcription factors within their promoter regions. To shed light on the combined effects of multiple transcription factors in gene regulation, we used a synthetic and computational biology approach to construct and characterize dual-input gene regulatory networks in *Saccharomyces cerevisiae*. Inducible transcription factors, based on the GEV and Tet systems, were tested against a set of synthetic promoters with varying numbers of DNA binding sites. Expression of a GFP reporter in various concentrations of the inducers, estradiol and anhydrotetracycline, was used to quantify promoter activity and fit mathematical models. Understanding the effects of multiple transcriptional regulators will aid in the construction of predictable, large scale synthetic networks and provide a better understanding of natural gene regulatory networks.

Using a Biosensor Controlled Genetic Feedback Loop to Improve Microbial Biofuel Tolerance

Lindle, J.,
Dunlop, M., The University of Vermont

Advanced biofuels offer a promising alternative fuel source to gasoline and diesel because they are renewable sources of energy that are compatible with existing fuel infrastructures. Engineered microbes can synthesize these biofuels but can only tolerate limited concentrations in their environment. To increase biofuel yield, and to make biofuels an economically competitive fuel source, microbial biofuel tolerance must be increased. Previous research has shown that efflux pumps are an effective tolerance mechanism for cell growth in biofuels. These pumps are complexes of proteins that identify harmful compounds and remove them from the cell. However, overexpression of these pumps can also inhibit cell growth and, therefore, decrease biofuel production. Biofuel and pump expression toxicity must be balanced with each other in order to maximize cell growth.

Our research goal is to design a synthetic feedback loop that will be able to detect biofuel with a sensor protein, which will then trigger the expression of the efflux pump in the host *E. coli.* The protein MexR, native to *Pseudomonas aeruginosa*, was chosen as the biosensor because it detects oxidative stress, such as that caused by certain biofuels, in cells. In the feedback loop, MexR represses the expression of the efflux pump by binding to synthetic promoter regions. To best control the system we have chosen to use an alternative sigma factor, rather than relying on *E. coli*'s native ones which can respond to environmental stress, such as oxidative stress. A sigma factor native to *Pseudomonas putida* was chosen because studies show that it did not interact with two of *E. coli*'s main sigma factor families. In order to choose the best synthetic promoter region, we have designed a competition experiment for over 500 promoter regions, to allow our system to thrive in a dynamic environment.

Novel Combinations of Efflux Pumps Improve Fitness Landscape of E. coli Under Biofuel Stress

Turner, W.,
Dunlop, M., The University of Vermont

Novel combinations of efflux pumps improve fitness landscape of *E. coli* under biofuel stress
Recent advances in synthetic biology have enabled the construction of non-native metabolic pathways for production of next-generation biofuels in microbes. One such biofuel is the jet-fuel precursor Î±-pinene, which can be processed into high-energy pinene dimers. However, accumulation of toxic biofuels in the growth medium limits the possible fuel yield. Overexpression of transporter proteins such as efflux pumps can increase tolerance to biofuels by pumping them out of the cell, thus improving fuel yields. However, too many active efflux pumps can compromise the cell as well, creating a trade-off between biofuel toxicity and pump toxicity. Previous research has suggested that certain combinations of efflux pumps can confer additional tolerance compared to the individual pumps themselves. However, the functional form of the combination of the tolerance provided by each pump and the toxicity due to their simultaneous activity is unknown. Using differential equations, we have developed a growth model incorporating the trade-offs between toxicity of Î±-pinene and efflux pump activity to describe the dynamics of bacterial growth under these conditions. By analyzing biofuel toxicity and the effects of each efflux pump independently through a series of experiments and mathematical models, we propose a functional form for their combined effect on growth rate. The maximum growth rate as a function of promoter activity and biofuel concentration is modeled and compared to experimental data with the ultimate goal of guiding design of tolerance mechanisms to improve overall biofuel yield.

Crystal Structure of the E. coli AcpP=FabB Complex

Jackson, D., University of California, Irvine

The acyl carrier protein (ACP) plays a central role in both primary and secondary metabolism. In *E. coli*, the ACP AcpP is responsible for transporting a growing fatty acid during biosynthesis. In different biosynthetic pathways including fatty acid biosynthesis (FA), AcpP interacts with at least 20 enzymes while it delivers intermediates into different active sites. The interactions between AcpP and its partner enzymes are transient, which has hampered efforts to gain structural knowledge of how AcpP interacts with its partner enzymes. In order to gain a better understanding of how AcpP interacts with the ketosynthase (KS) FabB during *E. coli* FA biosynthesis, an enzyme-specific mechanism-based crosslinking strategy was used. By applying this method, a stable crosslinked AcpP=FabB complex was purified, crystallized, and the crystal structure determined to 2.4 Å. The AcpP=FabB structure is the first structure of an ACP/KS complex and reveals important residues for complex formation. This work represents a major breakthrough in understanding type II FA biosynthesis, and provides a structural basis for engineering the production of pharmaceuticals and biofuels in *E. coli.*

Engineering Cellulose-Degrading Complexes from Anaerobic Gut Fungi
Haitjema, C., University of California, Santa Barbara
Solomon, K. V., University of California, Santa Barbara
O'Malley, M. A., University of California, Santa Barbara

Title: Engineering cellulose-degrading complexes from anaerobic gut fungi

Charles Haitjema, Kevin Solomon, and Michelle O'Malley

Department of Chemical Engineering, University of California, Santa Barbara
Anaerobic gut fungi are the primary colonizers and decomposers of plant biomass in monogastric herbivores and ruminants. Their ability to hydrolyze complex lignocellulosic materials is of great promise to cellulosic bioenergy development, since enzymatic breakdown of biomass is a stringent bottleneck. It is hypothesized that large cellulolytic complexes, or cellulosomes, mediate cellulolytic activity within gut fungi. While bacterial cellulosomes are well characterized, the molecular architecture of fungal cellulosomes is poorly understood. We characterized the composition and structure of fungal cellulosomes from the gut fungal isolate *Piromyces sp. finn.* To do so, we developed a practical approach to isolate cellulosome components from fungal cultures under a variety of substrate growth conditions. Following the purification of fungal cellulosomes, we combined tandem mass spectroscopy with strand-specific transcriptomic sequence information to identify the key components of these cellulosomes. To our knowledge, these studies are the first to reveal the entire enzymatic ensemble that comprises fungal cellulosomes. In addition, we used size-exclusion chromatography to reveal complex formation of enzymes within cellulosomes, and we provide evidence that most dockerin-fused enzymes bind to high molecular weight proteins as putative scaffolding molecules, which have long remained elusive for fungal cellulosomes. Finally, we are employing these techniques to study and compare cellulosomes from other unique isolates of gut fungi. This information will be valuable for developing consolidated bioprocessing platforms to convert plant biomass to energy.

Highly Modular "Bow-Tie" Gene Circuits with Programmable Dynamic Behavior

Prochazka, L., ETH Zürich
Angelici, B., ETH Zürich
Häfliger, B., ETH Zürich
Benenson, Y., Harvard University

Despite synthetic biologists' best efforts, many synthetic gene circuits operate only when their components â?? the controlling inputs, the input-processing module, and the outputs â?? are extensively mutually optimized. Implementing flexible, programmable tools to control and reprogram living systems is still difficult. A possible solution lies in the application of the "bow-tie" network architecture. This architecture stipulates a focal component - a "knot" - mediating between the input processing module and the outputs, simplifying swapping/adding components, and introducing additional layer of control. Here we show synthetic bow-tie circuits that transduce multiple microRNA inputs into multiple protein outputs in cultured human cells, such that both the circuit logic and output dynamics are set independently and in parallel. Output dynamics are adjusted via two different knot configurations: the first being a transcriptional activator resulting in outputs reversibly tracking changes in the inputs; and the second comprising a recombinase-based cascade, transducing transient inputs into irreversible actuation. We experimentally constructed and extensively characterized bow-tie circuits in HEK293 cells, confirming their modularity, flexibility and scalability. We also validated the system using endogenously-expressed miRNA as inputs in additional cell lines. This novel platform can be used for biotechnological and biomedical applications in cell culture, model organisms and potentially in human therapy.

Enhanced Gene Disruption at Specific Promoter Region By Simultaneous Digestion of ZFN or Crispr/Cas System

Nomura, W.,
Masuda, A.,
Tamamura, H.,

Artificial zinc finger proteins (ZFPs) consist of Cys_2-His_2-type modules composed of about 30 amino acids with a ββα structure that coordinates a zinc ion. ZFPs that recognize specific DNA target sequences can substitute for the binding domains of enzymes that act on DNA to create designer enzymes, such as nucleases, recombinases [1], and methylases [2], with programmable sequence specificity. Genome editing and modification by the enzymes could be applied for many fields of basic research and medicine. Recent findings of new platforms for genome editing such as transcriptional activator-like effector (TALE) and CRISPR/Cas system expands the possibility of the technologies. These technologies enable efficient knock-out of gene of interest or regulation of gene function. In this study, ZFN pairs targeting promoter region of human telomerase reverse transcriptase (hTERT) were designed and constructed. Endogenous gene targeting by the paired ZFN showed successful induction of mutation at the upstream- and downstream-site of core promoter region of hTERT. The simultaneous digestion by both pairs of ZFNs showed enhanced disruption of the target gene. Guide RNAs of CRISPR/Cas system targeting the same sequence with ZFN were also constructed and tested. The enhancement of gene disruption by the simultaneous digestion at the upstream- and downstream-sites was observed as shown in the ZFN experiments. Cells with gene disruption at the core promoter region of hTERT showed decreased expression of hTERT. The study will provide the information for the role of hTERT promoter activity related to cell-specific phenotypes such as immortality of cancer cells. Furthermore, the gene targeting for specific promoter region could be applied for studies in Synthetic Biology.

[1] Nomura, W., Masuda, A., Ohba, K., Urabe, A., Ito, N., Ryo, A., Yamamoto, N., Tamamura, H. *Biochemistry*, (2012) **51**, 1510-1517.
[2] Nomura, W., Barbas, C. F., III *J. Am. Chem. Soc.,* (2007) **129**, 8676-8677.

Near-Perfect Digital Switching in a Synthetic Biosensor Circuit Achieved through Temporal Control of Circuit's Genetic Makeup

Lapique, N., ETHZ
Benenson, Y., Harvard University

Near-perfect digital switching in a synthetic biosensor circuit achieved through temporal control of circuit's genetic makeup

Nicolas Lapique1 and Yaakov Benenson1*

1Department of Biosystems Science and Engineering, Swiss Institute of

Technology (ETH Zurich), Mattestrasse 26, Basel 4058 Switzerland

*To whom correspondence should be addressed: kobi.benenson@bsse.ethz.ch

ABSTRACT
Biosensing of endogenous molecular inputs is widely used in basic research and applications that range from bioproduction to medicine. Synthetic circuits applications can be safe and reliable only if a high sensitivity is achieved without sensor leakage in the Off state. High dynamic range with such sensors has only been achieved by exogenous ligand IPTG, but for endogenous microRNA inputs the range has been modest. Here we dramatically improve sensor dynamic range and virtually reduce leakage in the Off state to zero via temporal control of sensor genetic makeup. We engineer a delay in output gene availability relative to the repressor module using site-specific recombinase Cre or a cascade of Flp and Cre recombinases, and measure dynamic range of up to 2000-fold. A considerable reduction of the leakage is achieved without loss of sensitivity because the strategy does not increase the strength of the repressor module and therefore it does not result in concomitant decrease of the On. The near-perfect digital biosensor we engineer can lead to safe medical applications, enable reliable biosensing, and facilitate the construction of large synthetic gene networks.

Parasight V1.0 – a Synthetic Biology Approach for the Detection of Parasites

Webb, A. J., Imperial College London
Kelwick, R., Imperial College London
McKeown, C., Imperial College London
Jensen, K., Imperial College London
Kylilis, N., Imperial College London
Baldwin, G. S., Imperial College London
Ellis, T., Imperial College London
Freemont, P. S., Imperial College London

One of the primary goals of synthetic biology is the application-driven generation of new parts, circuits, and systems to solve problems that as yet have not been adequately addressed. The parasitic infection Schistosomiasis affects over 200 million people worldwide, with estimates suggesting that a further 780 million people are at risk of infection. The causative agents are fluke worms of the *Schistosoma* genus, and infection only occurs when the cercarial larvae are able to penetrate the skin. To facilitate this, the cercariae secrete an elastase possessing a defined substrate specificity. Our project takes advantage of this property of the cercarial elastase, and has used it to design and create biosensors that are specific in targeting *Schistosoma.* The design of our biosensors is based on a two-pronged approach, 1) an accurate detection system which is targeted to the cercarial elastase, and 2) an easily measurable output. Using synthetic biology approaches we have engineered the biosensors to be "housed" in two bacterial chassis, *Bacillus subtilis* and *Escherichia coli.* The *B. subtilis* biosensors comprise a fusion protein that possesses the cell wall binding domain of LytC along with our biosensor component, whilst the *E. coli* biosensors are composed of the CmpX protein, which traverses the outer membrane, fused with our biosensor component. Both biosensor components are presented to the external environment, and possess cercarial elastase recognition motifs, as well as either a streptavidin binding peptide for the *E. coli* biosensors, which can be detected by a streptavidin fluorophore conjugate, or a poly-histidine tag for the *B. subtilis* biosensors, which can be detected using a poly-histidine tag antibody fluorophore conjugate. Thus, when the elastase is present it cleaves the biosensor at the recognition motif site, thereby releasing the detectable components. Here we report and discuss the progress we have made in this important project and share some exciting data.

Marinobacter Aquaeolei VT8 As a Model for the Production of High-Value Neutral Lipids in Bacteria

Barney, B. M., University of Minnesota

The study of biological pathways utilized to produce and accumulate energy storage compounds such as fats and oils has tremendous importance towards pursuing new routes in the production of sustainable fuels and new bioproducts to meet growing energy and production demands. Wax esters, the product of the esterification of a fatty acid and a fatty alcohol, are a highly valued class of neutral lipid utilized in cosmetics, pharmaceuticals, as fine lubricants, and potentially as fuels. Most commercial wax esters are synthesized from fossil fuels, and thus there is a growing interest in developing biological systems for the sustainable production of wax esters. Our laboratory has evaluated several strains of bacteria, including *Marinobacter*, *Acinetobacter*, *Rhodococcus* and *Psychrobacter* as potential models for both whole-cell systems analysis, and also as sources for enzymes for *in vitro* laboratory studies, and based on a number of factors, have chosen *Marinobacter aquaeolei* VT8 as an ideal model system for the study of wax ester production in a bacterium. Our characterization of key enzymes from this species have found enzyme activities that are several orders of magnitude higher than previously characterized enzymes from other sources, and have also revealed key kinetic features of these enzymes that would be pertinent factors to consider in biosynthetic design approaches. Theoretical models of these enzymes have enabled us to probe potential residues surrounding the active sites, and revealed key residues with a role in determining substrate specificity. Whole cell analysis, including RNA-seq studies of changes in transcription during the wax ester accumulation and degradation curve provide a unique look at how this bacterium regulates the production of genes involved in the wax ester production pathway, providing clues for potential efforts to improve lipid production in a bacterium that has evolved to naturally accumulate a neutral lipid. Recent results from enzyme studies, gene deletion experiments and transcriptional profiling of the entire genome are presented.

Engineering Autonomous Recombinase Switches

Borkowski, O., Stanford University
Subsoontorn, P., Stanford University
Endy, D., Stanford University

Synthetic modules are both under control of specific regulations which are the result of rational design and of the cell physiology which is the result of growth conditions. In bacteria growth-phase-dependent replication, transcription and translation machineries affect proteins productions and thus, synthetic module behaviors. We decided to use the impact of the cell physiology on protein concentration to control a synthetic module, the Recombinase Addressable Data (RAD), without the need of any costly inducer. We developed several strategies mainly based on a control at the transcriptional level to modify, with growth phases, the production of proteins (integrase and/or excisionase) involved in the RAD module. This growth-phase-dependence can be used as a new and economic tool for research and industrial production of molecules of interest.

Crispr-Cas9 Assisted Recombineering in Lactobacillus Reuteri

van Pijkeren, J. P.,
Oh, J. H.,

ABSTRACT

Clustered regularly interspaced palindromic repeats (CRISPR) and the CRISPR-associated (Cas) nuclease protect bacteria and archeae from foreign DNA by site-specific cleavage of incoming DNA. Type-II CRISPR-Cas systems, such as the *Streptococcus pyogenes* CRISPR-Cas9 system, can be adapted such that Cas9 can be guided to a user-defined site in the chromosome to introduce double-stranded breaks. Here we have developed and optimized CRISPR-Cas9 function in the lactic acid bacterium *Lactobacillus reuteri* ATCC PTA 6475. We established proof-of-concept showing that CRISPR-Cas9 selection combined with single-stranded DNA (ssDNA) recombineering is a realistic approach to identify at high efficiencies edited cells in a lactic acid bacterium. We show for three independent targets that subtle changes in the bacterial genome can be recovered at efficiencies ranging from 90-100%. By combining CRISPR-Cas9 and recombineering we successfully applied codon saturation mutagenesis in the *L. reuteri* chromosome. Also, CRISPR-Cas9 selection is critical to identify low-efficiency events such as oligonucleotide-mediated chromosome deletions. This also means that CRISPR-Cas9 selection will allow identification of recombinant cells in bacteria with low recombineering efficiencies, eliminating the need for ssDNA recombineering optimization procedures. We envision that CRISPR-Cas genome editing has the potential to change the landscape of genome editing in lactic acid bacteria, and other Gram-positive bacteria.

Library of Azotobacter Vinelandii Identifies a Strain with Biofertilizer Potential

Eberhart, L. J., University of Minnesota
Barney, B. M., University of Minnesota

Nitrogen is the most expensive nutrient input for photosynthesis-dependent crops. Nitrogen fertilizers that are generated through current industrial methods are both economically and environmentally costly due to the large amounts of fossil fuels consumed during production and transportation. This research aims to develop a novel means to make nitrogen fertilizers by engineering bacteria to extracellularly produce excess nitrogen to support the growth of algae or other photosynthetic species. *Azotobacter vinelandii* naturally fixes atmospheric nitrogen and grows under aerobic conditions, making it an ideal bacterium for use as a renewable biofertilizer to support crop growth.

Transposon mutagenesis was used to generate random insertion mutants of *A. vinelandii* to identify genes associated with a nitrogen-producing phenotype. Strains of interest were selected by growing them on agar plates containing no nitrogen along with a nitrogen biosensor strain that is only able to grow in the presence of an exogenous nitrogen source. Using these high-throughput methods to screen the library of *A. vinelandii*, six mutants were identified that enable the biosensor strain to grow. Each of these mutants had a transposon insertion at various locations within a specific gene. Site-directed deletion of this gene in *A. vinelandii* also resulted in a strain that supports the growth of the biosensor strain, confirming that disruption of this specific gene is responsible for the nitrogen-producing phenotype. When grown in liquid culture or on agar media where no other nitrogen source is present, this gene deletion strain, AZBB109, also enables the growth of various algae strains that lack the ability to fix nitrogen on their own.

Efforts are underway to determine the nitrogen compound responsible for the nitrogen production phenotype in this gene disruption. Regardless, strain AZBB109 has potential as a biofertilizer to sustain algal cultures, and prospective applications in other important agricultural crops. This research shows that these screening methods can successfully identified an *A. vinelandii* strain capable of producing nitrogen at levels that allow various strains of algae to grow under conditions that they otherwise could not. Future experiments will focus on discovering additional gene manipulations in *A. vinelandii* that further enhance its level of extracellular nitrogen production.

Effective Dynamic Modeling of Transcriptional RNA Circuitry in Cell-Free Transcription-Translation (TX-TL) Systems

Hu, C., Cornell University
Varner, J. D., Cornell University
Lucks, J. B., Cornell University

RNA genetic circuitry is emerging as a powerful tool to dynamically control gene expression, and RNA networks of increasing complexity are beginning to be demonstrated. However, little work has been done to create a theoretical foundation for RNA network design. A prerequisite to this is a quantitative modeling framework that accurately captures the dynamics of RNA networks. In this work, we set out to develop an effective dynamical model of RNA transcriptional circuitry using our recent work on measuring the dynamics of an RNA transcriptional cascade using cell-free transcription-translation (TX-TL) reactions as a test case. For simplicity, we choose as a starting point the minimal number of ordinary differential equations that model the basic processes of RNA transcription and degradation and the translation of a fluorescent reporter protein. As a framework for obtaining parameters for this model, we use a model identification strategy that takes into account the full time course of dynamic experiments to design a sequential set of TX-TL experiments to perform. We show that all 15 parameters of this model can be determined from only four simple experiments. Interestingly, we show that inconsistencies between model predictions and experimental time trajectories allowed us to discover a previously unknown maturation step required for proper RNA regulator function, demonstrating that our approach can lead to model-guided discovery of basic mechanistic features of regulator function. We conclude with a description of how our model can be used to predict the dynamics of new RNA networks, and how parameter sensitivity analysis will allow us to identify the most important features for dynamic tuning in RNA network design.

Combinatorial Assembly of Large Biochemical Pathways into Yeast Chromosomes for Improved Production of Value-Added Compounds

Yuan, J., National University of Singapore
Ching, C. B., Singapore Institute of Technology

Saccharomyces cerevisiae as a eukaryotic organism is particularly suitable as microbial cell factory because it has interesting features like membrane environments for supporting membrane-associated enzymes and its capability for post-translational modifications of enzymes from plants. However, *S. cerevisiae* does not readily express polycistronic transcriptional units, which represents a significant challenge for constructing large biochemical pathways in budding yeast. In the present study, we developed a novel approach for rapid construction of large biochemical pathways into yeast chromosomes. Our approach takes advantage of antibiotic selection for combinatorial assembly of large pathways into the δ-sites of retrotransposon elements of yeast chromosomes. As proof-of-principle, a five-gene isobutanol pathway and an eight-gene mevalonate pathway were successfully assembled into yeast chromosomes in one-step fashion. To our knowledge, this is the first report to exploit δ-integration coupled with antibiotic selection for rapid assembly of large biochemical pathways in budding yeast. We envision our new approach could serve as a generalized technique for large pathway construction in yeast – a method that would be of significant interest to the synthetic biology community.

Models of the Cell for Synthetic Gene Circuits Design
Poh, C. L., Nanyang Technological University
Jayaraman, P. K.,

Models of the Cell for Synthetic Gene Circuits Design

Premkumar Jayaraman, **Chueh Loo Poh**

School of Chemical and Biomedical Engineering, Nanyang Technological University, Singapore

Abstract
Synthetic Biology is an emerging and rapidly developing field that applies engineering principles to biology to systematically engineer biological systems with new functions that, for example, address important and challenging problems related to healthcare, energy and the environment. A challenge in Synthetic Biology is to achieve predictable and efficient engineering of the biological systems. This challenge is largely due to the complexity of biological systems. To engineer biological systems with new functions, foreign genetic circuits are often designed and introduced into microorganism (host) to achieve the desired functions. Consequently, this adds burden to the host as these foreign circuits would draw cellular resources and energy from the organism. This usually affects the growth of the host which will in turn affect the performance of the genetic circuits. It will be useful to be able to study this effect *in silico* before actual wet laboratory experiments. To this end, it is important to have models that are able to predict host-dependent circuit functions and burden on the host. The insights from those models could then be used to optimize the gene circuit and reduce the host-burden. In this talk, we will present models of *Escherichia coli* cell to interpret and predict the behaviour of synthetic gene circuits and its host-effect. We have verified the modelâ??s ability to mimic the wild-type cell state by estimating the growth reduction under different external glucose levels. We also validated that the model is able to predict the effects of transcription and translation inhibiting antibiotics on cellular growth under limiting nutrient conditions. After verifying the model without foreign gene circuits (i.e. wild-type), we carried out experiments to verify the models when foreign gene circuits were introduced. Our results show that our model can quantitatively account for the impact on cellular growth rates when simple constitutive and inducible devices are introduced.

Systematic comparison of enzymatic error correction methods using deep sequencing

Zhang, D.,
Lubock, N. B.,
Church, G. M., Harvard University
Kosuri, S., Harvard University

Advances in *de novo* gene and genome synthesis will broadly benefit biology and biotechnology. Currently, one of the main bottlenecks for accurate, affordable, and high-throughput gene synthesis is the frequent occurrence of errors in the synthesized DNA sequence. Although various enzymatic error correction methods have been developed to address this problem, statistically effective evaluation and comparison of methods is difficult. Here we couple error-enriched template synthesis with next-generation sequencing to benchmark most previously reported enzymatic error correction methods. We report the landscape of synthesis errors before and after enzymatic corrections and compare the performance of each method in correcting deletions, insertions, and mismatches. This represents the first study that leverages deep sequencing to characterize error correction methods on a scale that is orders of magnitude larger than any past report and provides a method to benchmark novel enzymatic error correction techniques.

Engineering Yeast as a Host for Modular Polyketide Production

Kolesnikov, D., Stanford University
St Onge, R. P., Stanford University
Davis, R., Stanford Genome Technology Center

Natural products have demonstrated incredible value to human welfare through their pharmaceutical properties. Polyketides are a particularly interesting class of natural products, however, these compounds are often difficult to synthesize and their hosts are rarely amenable to genetic engineering. We are interested in polyketides produced by bacterial polyketide synthases (PKSs). These modular enzymes assemble their products in an assembly-line fashion, and module-prediction at the DNA level corresponds to the specific chemical modifications that the enzyme performs on its growing poly-acyl chain. We seek to express these proteins and their natural products in yeast, and ultimately to engineer novel polyketides by "swapping" modules at the DNA level, resulting in novel natural products.

Unfortunately, the expression of these large, heterologous proteins remains challenging. Preliminary work has shown that codon optimization significantly improves protein expression. We have partnered with Gen9, a DNA synthesis company, to assemble fully codon-optimized bacterial PKSs from the disorazole, nystatin, and amphotericin biosynthetic clusters. We have also developed C-terminally fused selectable markers (*URA3, HIS3*) for the purposes of selecting for the production of our proteins of interest. By fusing a bacterial PKS gene to a selectable marker and introducing the construct into the yeast deletion library (pilot experiments resulted in 85% coverage), we can identify strains which improve heterologous protein expression. These hits will be verified in individual growth experiments, and eventually combined into a yeast "superhost," which can then be used in future metabolic engineering projects.

Rational Design for Enzyme Engineering Using a Next Generation Gene Synthesis Approach to Generate Combinatorial Variant Libraries

Saaem, I., Gen9
Hudson, M.,
Kung, L.,
Schindler, D.,
Guido, N.,
Leake, D., Dharmacon Products

ACS Synthetic Biology

Rational Design for Enzyme Engineering Using a Next Generation Gene Synthesis Approach to Generate Combinatorial Variant Libraries

Journal:	ACS Synthetic Biology
Manuscript ID:	Draft
Manuscript Type:	Viewpoint
Date Submitted by the Author:	n/a
Complete List of Authors:	Saaem, Ishtiaq; Gen9, Inc., Hudson, Michael; Gen9, Inc., Kung, Li; Gen9, Inc., Schindler, Daniel; Gen9, Inc., Guido, Nicholas; Gen9, Inc., Leake, Devin; Gen9, Inc.,

Rational Design for Enzyme Engineering Using a Next Generation Gene Synthesis

Approach to Generate

Combinatorial Variant Libraries

Ali Saaem, Mike Hudson, Li Kung , Dan Schindler, Nick Guido, Devin Leake

Gen9, Inc., 840 Memorial Drive, Cambridge, MA 02139

The need for novel enzymes with a wide-range of activities has dramatically increased recently, and enzyme engineering strategies are evolving with innovative ways for creating combinatorial DNA libraries, advanced HTS strategies, and computational design approaches. The challenge for successful enzyme engineering is often narrowing the exploration of sequence space from the impractical size of all possible sequence permutations to a manageable number of variants. To increase the rate of success, researchers generate libraries that target specific regions of the target enzyme, or even individual positions, and create variants within these regions or positions. Designed libraries therefore aim at increasing the ratio of advantageous to deleterious mutations, increasing the â??smart-nessâ?□ and utility of the library. One method for generating this type of rationally designed library uses next generation gene

23 synthesis. With the BioFabÂ® process as a starting point, we have developed a method that allows for a

24 rational design of high variant combinatorial libraries to be used for enzyme engineering. This unique

26 approach to the creation of high variant libraries shows a number of advantages to traditional

27 approaches, including a lack of representation bias. Here, we show that using a next generation gene

28 synthesis approach we are able to very rapidly create highly specific and diverse libraries.

ACS Paragon Plus Environment

Optogenetic Characterization Methods Overcome Key Challenges in Synthetic Biology
Tabor, J., UC San Francisco

OPTOGENETIC CHARACTERIZATION METHODS OVERCOME KEY CHALLENGES IN SYNTHETIC BIOLOGY

Jeffrey J. Tabor

Rice University, 6100 Main Street, Houston, TX, USA, 77019; E-mail: jeff.tabor@rice.edu
The goal of synthetic biology is to understand how to program systems-level biological processes, such as the growth of an artificial tissue, by writing unnatural DNA sequences. In order to gain better understanding of, and control over, the path from DNA to cell- and organism-level processes, synthetic biologists have adapted a modular design framework from electrical and systems engineering. In the â??biological systems engineeringâ?? framework, the manner in which individual genetic components, such as signaling proteins or transcription factors, transduce, transform, and transmit biological signals is first understood through rigorous characterization. The goal is then that components can be assembled into higher-order devices, and eventually systems, whose performance can be predicted from the properties of the components. Though researchers have made notable progress, the actual utility of the biological systems engineering framework has been limited by an inability to directly characterize the dynamical performance features of biological components, and thus predict how they will behave when composed and deployed in different contexts. The result is that synthetic biology is not yet a mature engineering discipline. To address this limitation, we have recently developed an all-optical biological â??function generator and oscilloscopeâ?? framework that allows us to directly characterize the dynamical signal processing properties of genetic components in the native cellular environment. In particular, we have created sine waves and linear ramps of a transcriptional repressor in E. coli and shown that the promoter that it regulates transforms the repressor signal linearly with a seven-minute time delay. Our method is based upon our previously engineered light-switchable bacterial two component systems, mathematical models of their input/output dynamics, computational algorithms to design light inputs to program custom gene expression dynamics, and custom-built optical hardware for growing cells in well defined light conditions and making precise fluorescent protein measurements. Ours, and related methods, can help overcome the major characterization bottlenecks that have limited synthetic biology, and along with future advances, enable a complete adaptation of the systems engineering framework to biology.

Sensitive detection of proteasomal degradation using orthogonal gene circuits

Zhao, W., Rice University
Zeng, Y.,
Segatori, L., Rice University

The ubiquitin proteasome system (UPS) has been linked to the development of a diverse range of diseases characterized by protein misexpression and accumulation of aberrant proteins. Evidence of the molecular mechanisms that could be potentially targeted to enhance UPS activity have begun to emerge, including the upregulation of UPS components and modulation of the conformational changes that are rate-limiting for peptidase activity. These studies clearly suggest that upregulation of UPS activity holds potential for the treatment of protein misfolding diseases in which the formation of nonnative proteinaceous inclusion bodies is associated with neurodegeneration. However, the molecular mechanisms controlling proteasomal activity are not sufficiently understood to rationally design compounds to upregulate UPS activity. Pharmacologic agents that enhance UPS activity are rare and hard to discover because current in vitro technologies
to screen compounds that enhance UPS activity fail to recapitulate the complexity of the quality control system within the context of a cell. Moreover, existing cellular assays are based on reporter proteins engineered to function as model UPS substrates, which are poorly suited to screen for enhancement of proteasomal degradation, since a loss of signal in these assays can arise from disruptions of many basic cellular processes.

To develop a mammalian cell-based platform to quantify proteasomal degradation that can overcome these challenges, we explored a series of orthogonal genetic circuits that interface with the natural proteasomal degradation machinery to convert modulation of UPS activity into an easily detectable fluorescent output signal. We developed a platform (Deg-On system) based on an engineered tetracycline repressor (TetR-Deg) that functions as UPS substrate and as repressor of GFP expression. As a result, TetR-Deg serves as a signal inverter that converts enhancement of proteasomal degradation (and thus reduced, potentially undetectable cellular concentration of a proteasome substrate) into increase in GFP output (Nature Communications, accepted). Guided by predictive modeling, we enhanced the circuit's signal sensitivity and dynamic range by introducing a feedback loop that enables self-amplification of TetR. By linking UPS activity to a simple and tunable fluorescence output, these genetic inverters will enable a variety of applications, including screening for UPS activating molecules and selecting for mammalian cells with different levels of proteasome activity.

To introduce memory of the output signal we are also currently exploring alternative circuit architectures, including an UPS sensitive toggle switch. We report here the use of synthetic biology tools for the design and implementation of tunable cell-based platforms to quantify the enhancement of proteasomal degradation in tissue culture that will be generally useful for discovering and characterizing natural and synthetic proteasome activators that target degradation in a wide range of cell types and for different therapeutic applications.

Genetic Reporter Systems for Understanding the MAR operon in Escherichia Coli

Rossi, N. A.,

Over the past decade, there has been an increasing amount of interest in bacterial individuality arising in clonal populations[1]. This emergent behavior often arises from stochasticity at the transcriptional level and is modulated by elements of the greater genetic network[2]. I seek to understand the role that interlocking negative and positive feedback loops within the resistance regulons of *Escherichia Coli* serve to maximize population fitness while maintaining phenotypic heterogeneity in individuals. To do this, I have developed a series of plasmid-based dynamic reporter systems to measure the expression of marA and downstream genes in real time. It is important to note that the dynamics of the genes must be normalized to basal fluctuations in total transcription – so-called extrinsic noise. Therefore these reporter systems have the capacity to filter out unwanted variation through cross-correlative analysis with consistently induced genes such as sigma factors.

This platform allows us to gather quantitative data on the pulsing of various proteins as a result of transcriptional stochasticity and genetic network architecture. By combining the reporter systems with iterative deletions in the surrounding genetic architecture, we can further our understanding of how each component gene contributes to emergent phenomena in genetic networks.

1. Davidson, C. J. & Surette, M. G. Individuality in Bacteria. *http://dx.doi.org/10.1146/annurev.genet.42.110807.091601* (2008). doi:10.1146/annurev.genet.42.110807.091601

2. Sanchez, A. & Golding, I. Genetic Determinants and Cellular Constraints in Noisy Gene Expression. *Science* **342,** 1188–1193 (2013).

Genome Wide Manipulation of Transcription by CRISPR-Cas9 Transcription Factors in Saccharomyces cerevisiae

Smith, J.,
Schlecht, U.,
Suresh, S.,
Aparicio, A. M.,
Steinmetz, L. M., European Molecular Biology Laboratory (EMBL)
Davis, R. W., Stanford University
Parts, L.,
St Onge, R. P., Stanford University

Bacterial type II CRISPR-Cas9 systems have been widely adapted for RNA-guided genome editing and transcription regulation in eukaryotic cells. We have optimized different activator and repressor fusions, and assessed the effects of guide position, to optimize Cas9-mediated transcriptional regulation in yeast. Using these data we developed a computational method that uses published data to select guides that activate or repress genes with superior specificity and potency. As a proof of concept we activated and repressed several known drug targets, investigated the effects of transcriptional changes on the protein-protein interaction network, and investigated the effects of guides targeting bidirectional promoters. We are currently applying these methods to answer a variety of questions on a genome-wide scale using competitive assays involving pools of strains expressing individual guide RNAs.

Modulation of gene expression via directed electrical signaling; voltage sensitive promoters

Okasheh, Y.,
Marshall, N. M.,
McBee, R.,

The ability to interface biology with electronic components is extremely limited, in large measure because cells do not readily recognize electronic inputs in a way that we understand. Electrical connectivity in biological systems is mediated by kinetically isolated sets of chemical mediators, including pools of flavins, nicotinamides, glutathione carbon sources and oxygen (for example). The key challenge towards achieving facile "communication" between biologics and electronics is finding ways in which electrical inputs from electronics can be converted into chemical signals that can be "understood" by cells. In this work, we aim to increase the toolbox of bioengineers by achieving direct electrical control of transcription in *E. coli* through organic dye (redox mediator) assisted electron transfer into cells. As the dyes exchange electrons directly with an electrode surface, they serve as a "switching point" to convert electrical input into a chemical signal, which can be sensed by the cell through binding to proteins or alterations of intracellular chemistries. Transcriptome analyses of *E. coli,* in the context of a microbial fuel cell, have been conducted and we have found several highly regulated genes of interest. The expression regulators upstream of these genes will be used in fluorescent protein expression plasmids for quantitative expression assays to validate electrical control of expression and test the efficacy of such a system in synthetic applications. By utilizing multiple redox mediators, we will also investigate "multiplexing" of our system, wherein multiple gene expression profiles could be altered simply be tuning the applied electrical potential. Achieving this would give us a digitally controlled expression system that can be used in bulk fermentation applications, among numerous other applications.

AUTHOR INDEX

AUTHOR INDEX

AUTHOR INDEX

AUTHOR INDEX